"This is the book my heart has been yearni[illegible] personal narrative and clear-eyed science. With their emphasis on cultivating community, tending to personal rejuvenation and mental well-being, and offering practical tools for lightening our footstep on our beloved Mother Earth, the authors lift up every value we can all embrace right now to not only pluck us from the brink of the cataclysm but also transform the soul of the world."

—Mirabai Starr, author of *Wild Mercy: Living the Fierce and Tender Wisdom of the Women Mystics*

"Simultaneously intimate and thoroughly researched, this gem of a manual shares the story and the wisdom of the Good Grief Network, the first organization to create collective space for processing climate distress and transforming it into generative action. Changing the systems that cause harm to people and the planet will require collective effort, to be sure, and this book outlines the inner practices that such an effort will require of us all."

—Sarah Jaquette Ray, Professor and Chair of Environmental Studies at Cal Poly Humboldt and author of *A Field Guide to Climate Anxiety: How to Keep Your Cool on a Warming Planet*

"It goes like this: we pass a cup, back and forth, from one to the other. Inside the cup is what we most need to drink to keep going in this long struggle—a potent brew of inspiration; wisdom; real, heartfelt humanity; stories; tears; and laughter. We drink, take a long sip, and go on to do what is ours to do. The doing refills the cup with hope. This book is that cup."

—Susanne Moser, PhD, coeditor of *Creating a Climate for Change* and founder of the Adaptive Mind Project

"*How to Live in a Chaotic Climate* offers a needed space to contend with the worries, uncertainties, and grief associated with the consequences of global climate change. It is inspirational and courageous in the call to seize the moment by creating communities, investing in each other, and working for meaningful social change."

—Brett Clark, coauthor of *The Robbery of Nature* and Professor of Sociology, University of Utah

HOW TO LIVE IN A CHAOTIC CLIMATE

10 STEPS TO RECONNECT WITH OURSELVES, OUR COMMUNITIES, AND OUR PLANET

LaUra Schmidt

with Aimee Lewis Reau & Chelsie Rivera

SHAMBHALA

Shambhala Publications, Inc.
2129 13th Street
Boulder, Colorado 80302
www.shambhala.com

© 2023 by LaUra Schmidt

"Funeral for a Future," from *Bury the Seed: Poems for Releasing More Life into You*, by Brooke McNamara, is reprinted with permission of the author. © 2020 by Brooke McNamara.
"Bedrock," by Kristan Klingelhofer, is reprinted with permission of the author. © 2023 by Kristan Klingelhofer.

Cover art: CPD-Lab/iStockphoto
Cover design: Katrina Noble
Interior design: Katrina Noble

All rights reserved. No part of this book may be reproduced in any form or by any means, electronic or mechanical, including photocopying, recording, or by any information storage and retrieval system, without permission in writing from the publisher.

9 8 7 6 5 4 3 2 1

First Edition
Printed in Canada

Shambhala Publications makes every effort to print on recycled, acid-free paper.
Shambhala Publications is distributed worldwide by Penguin Random House, Inc., and its subsidiaries.

LIBRARY OF CONGRESS CATALOGING-IN-PUBLICATION DATA
Names: Schmidt, LaUra (Co-founder of Good Grief Network), author. | Lewis Reau, Aimee (Co-founder of Good Grief Network), author. | Rivera, Chelsie, author.
Title: How to live in a chaotic climate: ten steps to reconnect with ourselves, our communities, and our planet / LaUra Schmidt with Aimee Lewis Reau and Chelsie Rivera.
Description: First edition. | Boulder, Colorado: Shambhala Publications, Inc., [2023]
Identifiers: LCCN 2022051660 | ISBN 9781611809930 (trade paperback)
Subjects: LCSH: Environmental psychology. | Climatic changes—Psychological aspects. | Climatic changes—Influence.
Classification: LCC BF353.5.C55 S36 2023 | DDC 155.9/15—dc23/eng/20221208
LC record available at https://lccn.loc.gov/2022051660

To Mackenzie, Caleb, Emma, and Jude

And to young people everywhere. May each of us practice courage and embody connection to cocreate life-centered futures for you and all beings.

CONTENTS

A LETTER FROM THE AUTHORS

Dear Brave One,

We do not have it all figured out. No one does. These times are full of unexpected twists and painful turns. But there is joy and meaning waiting to be uncovered and experienced. The story of these times is being written with each breath, each step, each seed planted. That has always been true, but it feels more salient in times of great tumult.

This book, and our 10 Steps to Resilience & Empowerment in a Chaotic Climate program, which the book is based on, is our attempt to share with you the wisdom we have gleaned along our personal journeys. We see ourselves as bridge builders and mash-up artists. By exploring a number of disciplines and modalities and funneling them through our life experiences, we have arrived here with the wisdom in these pages. We are not experts at anything but fellow seekers who stand on the shoulders of many giants. We have had a variety of teachers who have helped us shape these steps and our philosophies. Many of our teachers have been the people we have had the honor of sitting with in our 10-Step program, where each of us becomes a teacher by sharing the wisdom earned in our own journey.

Every day, we are living the process outlined here, and we are always working to advance the dialogue about how to live openheartedly in these disruptive times. This is an emerging conversation with no one right pathway forward. We invite you to keep the conversation—and your heart—open as we feel our way through these times and plant seeds for future paradigms that are life-centered. May meaning, joy, connection, and courage serve as guides along this uncertain path.

—Onward,
LaUra and Aimee

PREFACE

> This time represents a critical crossroads for humanity, a teetering point of choice that will determine the future of all life.
>
> —Sherri Mitchell, *Sacred Instructions: Indigenous Wisdom for Living Spirit-Based Change*

Before we begin, slow down. Take a deep breath. Collect yourself. Arrive here, fully.

We are at the top of a mountain, and we know we cannot go back the way we came.[1] Our descent is not illuminated or marked. There is no path. No one has been here before. It is unknown. It can feel scary. And we must move forward.

This is where we find ourselves right now, collectively, as we witness the unfolding of our biological and social systems. The dominant culture's social systems, institutions, and norms are endangering life on Earth, forcing us into a time of deconstruction and reordering. This moment in history holds both exciting possibilities and frightening unknowns. Right now, those of us living within the dominant culture are stuck between two worlds, in the liminal space situated between what was and what is to come. A new world is on the horizon, but to arrive there we have to descend the steep, rocky mountain and enter into the depths of the night. It is best if we go together.

As our planet warms, the human and biological systems with which we are familiar are unraveling, dropping us into a vast unknown. There has already been much suffering, and we do not know how long this night may stretch but denying or avoiding it does not make it go away. Like it or not, we are in it for the long haul.

But that does not mean we should simply give up or that our lives are over. Our reactions to the state of the world directly inform the types of lives we will lead and what types of futures we are capable of cultivating. When we deny reality and repress our emotions, we dull our life experiences as individuals and continue to do harm as a collective. But when we embrace the unknown and work cooperatively, we plant seeds that may grow into *new ways of being* for future generations. "New ways of being" is a general phrase that builds on the new and ancient story of "interbeing" that Thich Nhat Hanh taught about and Charles Eisenstein advances in his book *Climate: A New Story*. These ways of being are an embodied knowing that existence is relational, generative, and diverse. These new paradigms can incorporate our modern understanding of the world with ancient ways of being that were in right relationship with life and life-supporting systems. Other cultures have existed—and still exist—in mutual relationship with life and life-supporting systems. We are not advocating for one new way of being or for appropriating other cultures but for listening to and working with people who have been marginalized as we remember what a truly life-centered civilization might look like and begin reorienting toward cocreating it. By picking up this book, you have already begun to sow the seeds of new ways of being.

We will share many of the tools we have developed for the Good Grief Network (GGN), a global community established to help support people in metabolizing their heavy, painful feelings and creating personal and collective resilience to weather these tumultuous times. We believe that authentically bringing people together in times of great division is one of the most important salves for deep disconnection and fragmentation. There are a wide range of reasons that people come into our GGN circles. Many people of child-bearing age question how they can ethically bring children into such a broken world. Parents worry about the living conditions that will be available to their children and their children's children. Young adults contemplate meaning and purpose in a disordered world. Burned-out activists mourn that their work was not enough to change the tide. Climate-crisis survivors wonder how to move forward after watching their homes disappear in flames or under water. But mostly, people come into our circles because once awakened, they are riddled with isolation, fear, and grief about the times we are living in and over the future world they project.

Underneath these worries lie two common themes: uncertainty and tremendous grief. Our planet is experiencing an unraveling that humanity has never before seen. Some scientists say that we have crossed critical planetary turning points and that it is only a matter of time before our planet has moved out of a habitable range for humanity. Others say there is a sliver of time left, and we have to get to work, *right now*. In this moment, record-setting storms and wildfires, melting permafrost, and calving glaciers point to the reality that we have created instability in our climate. The sixth mass-extinction event reflects the breakdown of our biosphere. Ecosystems can only tolerate so much abuse before they're drastically transformed. Simultaneously, we are also witnessing the buckling of our deeply flawed social systems. Polarization and Othering seem to be at an all-time high. There is an ongoing threat of nuclear war. Rates of depression, especially among our youth, are growing.[2] And during the first two years of the pandemic, "the world's 10 richest men have doubled their fortunes, while over 160 million people are projected to have been pushed into poverty."[3]

Some believe we are on the brink of human extinction. Others feel we are on the frontier of building better ways of being, just in time to save humanity and countless other species. We echo the words of youth activist, Greta Thunberg, as she chides the politicians at a United Nations Climate Action Summit: "Change is coming, whether you like it or not."[4] As we live through this deep transition, we are tasked with minimizing suffering in the present and starting to cultivate and move toward new paradigms even if we do not arrive there. Instrumental to building a life-supporting future is the deconstruction of the deeply held ideas, beliefs, and norms that are not serving humanity and the larger world. The rapidly changing biosphere and crumbling of our social world challenges many of our collective and personal beliefs. While this may be uncomfortable for some, these changes hold promise, as many of our cultural norms and beliefs perpetuate the oppression of women, communities of color, those most impoverished, and our planet as a whole. If we are indeed on the frontier of building back better, we are still being asked to collectively descend the mountain to reassemble our worldviews and create truly life-centered ways of being.

This is a lot to grapple with for a single human psyche, body, and soul. In our day-to-day lives, how do we go on clocking in for our day jobs, hugging our loved ones, working toward equity and justice for all, and enjoying the sunshine when we are constantly confronted with ongoing loss?

This book is for those of us who want to build stronger inner capacities and community as we climb down the mountain together. No one has all the answers for what comes next or how best to navigate the descent. We are all figuring it out as we go. We do know a few things, though. First, the current social systems are set up to disproportionately impact those of us in the most vulnerable and marginalized groups, especially women, BIPOC (Black, Indigenous, and people of color) communities,[5] LGBTQIA+ folx, people with disabilities, and those with low incomes. We also know that there are real barriers to financial security, care, and time to rest for marginalized communities, meaning that as we descend down the mountain, the impacts of social and biological disruption will be borne most intensely by people in these demographics.

A major challenge of our time is learning to undo the cultural conditioning that allows us to perpetuate harm against these communities, and our larger Earth community, so that we can begin to see that we are all interconnected. Humans, by nature, are social and cooperative. We have a desire to help each other and see the best in each other. In a *Democracy Now!* interview some years ago, the activist and community organizer Grace Lee Boggs said, "The only way to survive is by taking care of one another, by re-creating our relationships to one another."[6] As we take steps down the rocky terrain, we must ask ourselves, *How can each of us extend care and minimize harm toward those most vulnerable?*

This is the time to be together, to break down the barriers that divide us. Exclusion and separation must end. We, as a common humanity, must be willing to live these questions together and learn to cultivate a relationship with an uncertain, dynamic, hurting planet Earth.

And so, instead of offering answers, this book offers an exploration of the following question: How can we live full and meaningful lives as the world changes and everything we thought we knew is challenged? It is a question we have been unpacking for years as we have created and facilitated 10-Step support groups designed to help build emotional intelligence and personal and community resilience for the long haul.

Regardless of whether you believe we are careening off the cliff toward our collective demise or on the brink of cocreating the cooperative, compassionate, loving, life-centered paradigms of our dreams, this book is for you—but only if you are ready to ask some hard questions about mainstream culture, about your perceptions of the world, and about yourself. Throughout these pages, you will not find clear-cut answers to every question. But you will find tools, ideas, and invitations for living openhearted in this liminal space between worlds. We offer wisdom for cultivating the most meaningful and joyous life right now because right now is all we have.

FUNERAL FOR A FUTURE

By Brooke McNamara

I held a funeral for a future
I had always thought was coming,
and buried the world's face as yet.

The silence then
turned me so tiny
the only way forward was to dream
downward

to an early day on earth
before a single heart beat.
The atmosphere filled
with an abiding, cataclysmic knowing —

that if everything
could be born,

every
thing

could
be
born.

Love promises no less.

But a future is gone now.
All we are is this.

Our way could be
to fall toward the medicine
seeded right inside
the untamable, fertile grief
remaking things.

Brooke McNamara is a Zen teacher, lineage holder, and poet. Her wise, piercing words get to the crux of our complex predicament with profound heart and courage. We chose to include this poem for its message that, out of the untamable, fertile grief, everything can be born. Our hope lies there.

HOW TO LIVE IN A CHAOTIC CLIMATE

Introduction

This book is largely based on the 10 Steps to Resilience & Empowerment in a Chaotic Climate program we, LaUra and Aimee, designed for the Good Grief Network (GGN). We have been running this program for a number of years, and it's proven helpful for folks experiencing burnout, grief, anxiety, fear, anger, or other uncomfortable feelings in the face of so many unraveling systems.

For us, it was firsthand exposure to both personal and community tragedy that forced us to confront the collective traumas around us. For you, it may be a communal trauma brought on by climate catastrophe, systemic racism, or other injustices that have personally affected you. Perhaps you're an activist answering the call to create a better society. Or maybe you're just trying not to lose your mind as you navigate day-to-day life in this wild, broken world.

You are not crazy. You are not alone.

Whatever your situation may be, the steps in this book are designed to help you find groundedness when it feels like the rug has been pulled out from beneath your feet. We work and then rework them in this order:

10 STEPS TO RESILIENCE AND EMPOWERMENT IN A CHAOTIC CLIMATE

1. Accept the Severity of the Predicament
2. Be with Uncertainty
3. Honor My Mortality and the Mortality of All
4. Do Inner Work
5. Develop Awareness of Biases and Perception
6. Practice Gratitude, Seek Beauty, and Create Connections
7. Take Breaks and Rest
8. Grieve the Harm I Have Caused
9. Show Up
10. Reinvest in Meaningful Efforts

The steps and exercises in this book will help you cultivate a relationship with uncertainty, nurture connections, and process heavy and painful emotions about the state of the world. You may have noticed that these steps aren't exclusively related to the climate crisis. That's because they can be used in a multitude of crises. There has already been so much suffering—floods, fires, droughts, and of course the pandemic, which have taken incalculable lives—and as we forge paths down the mountain, there will be many disruptions and more suffering.

In 2017, we used these steps to help support those who were shocked, confused, and saddened by the US presidential elections and the inauguration of Donald Trump. In the past few years, our sessions helped participants grieve the grave losses and lockdown during the COVID-19 pandemic, the war in Ukraine, and the worldwide wildfires and severe-storm impacts. Who knows what 2030 or 2050 will bring, but these steps will be here to offer support and help build connection through whatever we will face.

These steps promote *radical reconnection*, which can sustain us for the long haul. *Radical* is a word many of us hear often, perhaps because so many of us are longing to uncover the root of our dysfunctional ways. The term itself comes from the Latin *radix*, meaning "root." When we dig down deep enough, we can see that the root of many of these overlapping crises is a disconnection—from ourselves, from others, and from the more-than-human world.[1] This disconnection is a survival mechanism meant to protect us from pain, loss, and uncertainty, but it is terribly constrictive. Rather than remain constricted, we are encouraged to open up and practice vulnerability because it's the only way we can truly connect. In a dominant culture that normalizes the repression of our emotions and reinforces the systemic division by race, class, gender, and more, connection is a form of rebellion.

We want to encourage you to keep your heart open throughout the chaos—to reconnect when your fear responses tell you to shut down and get small. Reconnecting is arguably the most important thing you can do for our planet. It is a radical request, especially in a world where we've been conditioned to believe someone will save us: the politicians, the Food and Drug Administration, the Environmental Protection Agency, the United Nations, or maybe the charismatic activist. No one is coming to save us. We have to uncover our own personal power and courage. Each of us must look within to do the inner work of accessing our full range of feelings, cultivating our souls, and healing our individual traumas. We must also deconstruct harmful cultural narratives and learn to reopen when our fear responses have us shutting down. We must balance the inner work with the outer work of connecting to others and the world around us.

As we work to reconnect with ourselves, we can then begin to reconnect radically with people, animals, insects, trees, rivers, fungi, and the larger world around us. We can learn to trust others again, to create community, to do meaningful work in a world that can sometimes feel hopeless. We have often been taught that change happens from the top down, but what we have witnessed is that the folks at the top are laggards, falling behind the momentum from communities. When we call our legislators or go to protests, we are trying to convince the people in power to make a change. In a true grassroots movement, we at the community level have the power. We create the change. Instead of waiting for power and agency to be allocated to us, we use our superpowers to link up with other people, sharing wisdom and increasing our agency. Rob Hopkins, the cofounder of the Transition Town movement, offers, "If we wait for governments, it'll be too late; if we act as individuals, it'll be too little; but if we act as communities, it might just be enough, just in time."[2]

How you engage meaningfully in these times may look different for you than it does for any other person reading this book—and that's ideal. Meaningful action ought to be different for different people and for varied communities. We are moving away from homogeneity and embracing diversity on all levels. By reconnecting with yourself, you can uncover where your unique skills, passions, and experiences intersect. That is your place of true power as a heart-centered revolutionary. If your actions contribute to connection, healing, and growth, it is crucial work. For some, a reinvestment in parenting to create a more loving, conscious future is the work they are called to do. Or perhaps it looks like learning to grow organic, nourishing food that you can share with your community and takes a lesser toll on the soil. Maybe it looks like joining an organization that regenerates ecosystems. There are innumerable ways to reinvest into meaningful efforts (we'll explore this more in Step 10), and not all of them involve holding the megaphone in front of a crowd begging those in power to take the drastic action needed in these times. We aim to invite in creativity and nourish new ideas and meaningful effort.

Why Us? Why Now?

The creation of this book, like GGN more generally, is atypical. We are living in atypical times. This book was primarily written by LaUra. Due to ongoing health issues, Aimee had to take a giant step back. Chelsie Rivera, Aimee's friend from graduate school, stepped up. She helped us write and format the manuscript as well as represent Aimee's heart, stories, and voice in the book.

We, LaUra and Aimee, know a thing or two about being burned out, shut down, and operating from our survival responses. We built GGN together, based on a step program LaUra envisioned as part of her graduate studies in the environmental humanities—in particular, how people cope with uncomfortable feelings about the impacts of the climate emergency and the uncertain future of our planet. We created GGN not just because we saw the emotional toll that activism and collective grief were taking on our friends and colleagues but also because it was the program we needed to navigate our own healing process.

Long before our foray into activism or environmentalism, our life circumstances forced us to ask questions about suffering, reality, and the human condition. LaUra's early life was shaped by isolation and loss. After domestic violence prompted her mother to leave her father when she was five, LaUra and her sisters were caught in the crossfire of her parents' war. Both parents were psychologically absent, abusing substances, and addicted to stewing in their own misery. The systems meant to protect children from abusive parents failed LaUra and her sisters, leaving them to raise themselves as children. In high school, LaUra lost two of her closest friends within nine months of each other, forcing her to question the meaning of life being surrounded by so much suffering. As you might imagine, being a smart kid in a deeply dysfunctional family and experiencing world-shattering losses left LaUra unable to cope and without healthy support systems. She employed a variety of survival mechanisms to protect her from further pain. She shut her feelings down and coasted on autopilot for nearly a decade, investing in perfectionism and education as a means to transcend her life experiences. Yearning for more than the superficial life she was leading, LaUra continually looked to the wisdom of all types of teachers who had endured suffering in the past. And after college she found healing in the Adult Children of Alcoholics 12-step program, which later inspired the format for GGN's 10-Step program.

Meanwhile, Aimee's childhood looked like the American Dream. Raised by hardworking middle-class parents in a small Michigan town, Aimee was a straight-A student, an active member of the Catholic church, a dedicated soccer player, and a budding humanitarian. She spent much of her time in service and activism, from volunteer work to fighting for a living wage. But beneath the surface, Aimee was suffering from a variety of health conditions that began when she was a baby, including both chronic pain and undiagnosed ADHD. Rather than get to the root of her chronic illnesses, her family distracted themselves away from her pain and attention deficit, normalizing it among themselves and placing her on a variety of medications to cover up the symptoms. Wanting to be the peacekeeper in her family, Aimee learned to stuff away and downplay her feelings and suffering. Over time, she developed depression and began self-harming to cope. By her junior year of college, Aimee had been placed in a psychiatric hospital three times, with the doctors suggesting that her passion for activism was contributing to her mental illness and that she ought to reconsider her priorities.

When we met as undergraduate students, our connection was magnetic. We bonded over existentialism, music, the world's troubles, spirituality, and our own personal experiences of suffering. All of this is to say we have sat with, been crushed by, wrestled and bargained with, been seduced by, outrun and been outrun by, glossed over, and finally embraced grief—more than a few times. Between our formal education and the lessons we learned from the school of hard knocks, we have some wisdom to share about how to feel your way through the darkness that threatens to overwhelm.

Piecing together wisdom from ancient and contemporary thinkers, science, humanities, activism, therapeutic modalities, 12-step programs, losing loved ones, and love, the words on these pages are our offering to those who want to build personal and collective resilience as the world descends into chaos. The lessons we've learned are distilled into the steps in this book, as well as the support group program we've facilitated for thousands of participants through GGN over the years. These ten steps invite us into deeper self-understanding, an openness to connect to those we perceive as Other, to grieve our losses and also our complicity in the predicament, to rewild ourselves, and to identify our superpowers and engage in meaningful actions in a time when action can seem futile. These are the skills needed to endure this time of upheaval and to envision our next steps forward.

The Descent of the Dominant Culture

We use the terms *dominant culture* or *dominant paradigm* interchangeably and often throughout this book. By that we mean the set of expectations and norms that many of us in the postindustrial world abide by. The dominant culture is crafted and maintained by a power-over structure and perpetuates systems of exploitation and oppression by commodifying life and life-supporting systems (more on this later on). This way of being categorizes most things in binaries and is reductionistic in nature, leaving little room for complexity or the existence of multiple truths. Capitalism (more specifically, neoliberalism), patriarchy, racism, colonialism, and imperialism are all characteristic of the dominant culture. Though this way of being originated in the Western world, it has extended its tendrils worldwide. We founded GGN to help dismantle and diversify the dominant culture and begin healing the generations of damage it has caused. We do this by learning to live in dynamic relationship with the more-than-human world.

The systems upheld and perpetuated by the dominant paradigm have not served many of our fellow human beings. Generations of white patriarchy have silenced and murdered oppressed peoples—especially those who have spoken up to alarm the rest of us that something is terribly wrong with how we are operating on our planet. The dominant paradigm appears to work mostly in favor of a section of the population who are white, rich, cisgender, heterosexual, and male. And we would argue that it isn't actually working for these demographics either. Not only are the economic and industrial practices of the dominant paradigm destroying the very planet that nurtures life but the cultural norms perpetuated by this way of being thrive through separation, competition, alienation, and overconsumption. The dominant culture promotes lifestyles of isolation, leaving our survival mechanisms to deem many things as "threats," including people who do not look or think like us. Because of this, many of us live our day-to-day lives disconnected from each other, from the more-than-human world, and from our own emotions and embodied states. The truth is the dominant culture serves no one. By its very nature, it is antirelationship and antilife.

The dominant paradigm is so large and powerful that no one is untouched by its impacts. No matter how green or awakened or off-grid we strive to be, many of us participate in one way or another—whether by consuming animal products from concentrated animal-feeding operations, commuting to work with fossil fuels, buying goods that are not sustainably made, using single-use plastics, living on stolen Indigenous land, or perpetuating white-body supremacy.[3] The dominant paradigm and its stories live in us, and we perpetuate them. The dominant culture, and capitalism in particular, teaches us to compete with our peers instead of collaborating, leading to an epidemic of isolation and Othering where distrust prevails because everyone is seen as a threat. From a young age, many of us learn to put our feelings and needs aside so that we can compete and be more productive, more efficient, more successful. What is demanded of us is more, more, more, while our levels of fulfillment, meaning, and joy are less, less, less. Even in the most radical circles, these cultural narratives persist. Activists know this hustle well, as we so often work ourselves beyond the point of exhaustion and yet never feel we have done enough.

But it doesn't have to be this way. We may not be able to live entirely outside of the dominant paradigm or dismantle it single-handedly, but we can reclaim our personal agency by questioning and then deconstructing the stories we've been indoctrinated with. We often say that we don't believe in the dominant paradigm. Just like a fairy tale, it is a belief we can each opt out of. We are continuously trying to find new ways to withdraw from the dominant culture, and turn toward more generative, nourishing ways of being.

The Great Unraveling and the Long Dark

We are living in an age of unprecedented change. Our most seemingly predictable systems—ecological, cultural, economic, political, and more—are breaking down. By many thinkers, this breakdown is called the "Great Unraveling," a term popularized by Joanna Macy—an activist and scholar of Buddhism and systems thinking and the unofficial muse of GGN. The Great Unraveling is different from the fall of civilizations past. Unlike other cultural collapses in which a localized region or singular society is affected, the climate crisis is altering the entire globe, which then causes social collapse. This becomes a cycle: the social collapse hastens the biosphere collapse, and the biosphere collapse exacerbates the social collapse, and on and on.

As the world as we know it unravels, it reveals the interconnected problems of the climate emergency, rampant social justice issues, vast income inequality, and ecocide. At times, this change can feel exciting and hopeful. When old systems unravel, they carve out space for more life-centered systems to emerge. At other times, this systemic breakdown can feel scary or overwhelming as it increases uncertainty all around us. Messaging from the dominant culture conveys that uncertainty is bad, something to overcome. For privileged groups, the existing cultural and economic systems have mostly provided stability—or the illusion of it—with steady jobs, safe homes, a ladder for success, and enough money to pay for food, bills, vaccinations, and vacations. And even for those of us who do not benefit the most from the dominant culture, the idea that the climate emergency might change our way of life in unpredictable ways is anxiety provoking, to say the least.

As the Great Unraveling occurs, we're plummeted into a time of liminality, a period that is unrecognizable to us. Francis Weller, psychotherapist, writer, and soul activist, refers to this liminality as the "Long Dark."[4] We are not referring to the popular video game but a space where, under the veil of darkness, this rearrangement, reorganization, and culture building can happen. Weller explains, "We are entering the Long Dark. I use that term not negatively at all. I use it alchemically, that certain things can only happen in darkness. We are in a time of decay, a time of collapse, a time of endings, a time of sheddings. These are necessary."[5] Endings provide opportunities for new creations.

In this space, when we only have a flashlight's distance of visibility, we have many options available to us about how to proceed. Weller teaches us that this is a time to cultivate our souls, both individually and collectively, a time to move in the questions, the grief, the mystery. He invites us to ask ourselves and each other several questions during the Long Dark: "How do we become skillful in navigating our walk in the dark? How do we cultivate imagination? How do we cultivate collaboration? How do we cultivate fields of reciprocity with the Earth, within human and more-than-human communities, so that we're not extracting more than what can be replenished? How do we cultivate the spiritual values of restraint and mutuality?"[6] We don't know the duration of the Long Dark, but we do know that rich and meaningful lives are possible despite the looming question mark that is our future.

We can experience joy, love, and beauty on this planet, even as it changes around us. To do this, we have to build personal and collective resilience—an ability to find equanimity in unpredictable times and as the suffering around us increases. We do this not by avoiding the Long Dark but by facing it, moving with it. Being fully alive in this space requires that we experience our full range of feelings—excitement, playfulness, courage, peace, shock, fear, despair, rage, and grief—instead of being overwhelmed by them. Much of this book is focused on the practice of reembodying ourselves by feeling a wide range of emotions and reconnecting in generative ways so that we can feel joy more deeply, even in the face of the unknown future, and shoulder heavy feelings without collapsing under their weight.

The steps in this book are designed to help us parse out who we are, apart from the dominant culture, so that we can reconnect with ourselves, other beings, and the more-than-human world during the descent down the mountain. Connection has the power to ground us when the world is chaotic. Connection gives our lives meaning and offers joy, even in the dark. We can then invest ourselves into meaningful action—the kind that promotes relationship and regeneration. Meaningful action can be a salve for painful feelings like ecoanxiety, ecodistress, climate grief, and overwhelm because meaningful action isn't dependent on outcomes. We don't do this work because we believe that we, alone, can save our species and restore our planet. We do it because it's what needs to be done. It's generative work, and it fills us with purpose.

It also lays the groundwork for a *heart-centered revolution.* In this revolution, we center relationships, connectedness, and love in times of suffering and disconnection. We open to our interconnectedness with all beings and make decisions based on compassion and insight instead of egocentric motivations. The heart-centered revolution is brought about by our inner equanimity and our love for each other, ourselves, and our planet as a whole. Instead of thoughtless and selfish actions, we reinvest ourselves with an understanding of the consequences to the larger world.

The steps require a commitment to doing our own individual work and opening to the innate wisdom within each of us that has been covered up by the dominant paradigm.

The calling of the heart-centered revolution is to find opportunities to cultivate a truly just and life-centered world, even if we never see it come into existence.

A Framework for the Heart-Centered Revolution

> The greatest challenge of the day is: how to bring about a revolution of the heart, a revolution which has to start with each one of us?
>
> —Dorothy Day

In their book, *Active Hope: How to Face the Mess We're in without Going Crazy*, Joanna Macy and Chris Johnstone define three critical actions we must take to create a true heart-centered revolution: holding actions, creating new structures, and creating a shift in consciousness. While we may focus on one or two of these types of actions, all three can—and really must—be done simultaneously.

Protesting, signing petitions, writing legislators, passing bills, civil disobedience, and other forms of traditional activism are all *holding actions*—that is, actions that slow down the damage that the dominant culture wreaks on Earth. Holding actions are sacred and essential work. Holding actions are paramount to preventing the dominant culture from exterminating everything in its path. In recent years, we have witnessed the power of holding actions as the Black Lives Matter movement forced community leaders and politicians to finally start acknowledging the hundreds of years of systemic racism.[7] We have also seen pipelines shut down, and some species brought back from the brink of extinction. Holding actions are integral if we are to mitigate the absolute worst impacts of the climate emergency, ecocide, and social disruption. We must intensify and sustain our efforts to halt new fossil-fuel developments and stop carving up Earth, decimating wildlife, plant, and insect populations. We must protect those most vulnerable. Most of the work in the realm of activism is focused on this type of action. We also must create a new social and political order so that protecting life and life-supporting systems is the norm and not denied for being too radical.

And that leads to the next type of action: *structural change*. A true heart-centered revolution must reimagine the systems that (are supposed to) protect and support us and our planet, such as law enforcement, economics, politics, health care, energy, education, and so many others. The brokenness in our world cannot be healed until we start creating systems and structures that are truly life-centered. Our systems are reflections of our worldviews, and without a significant overhaul of our individual and collective values and beliefs, any structural changes will perpetuate the same harms that got us to the Great Unraveling.

All too often we stop at these first two, and that is where, on the individual level, burnout and resignation can happen. We get discouraged when we fail to see our hard work bring about the much-needed change we yearn for. We need to change our ways of thinking and being, too! For this reason, we are less concerned in this book about holding actions and the "how-tos" of reforming or breaking down the current, deeply broken systems. Instead, we are more concerned about freeing up our energy, ways of knowing, and imagination to begin focusing on what comes next. As the Great Unraveling continues and we compost that which no longer serves us through the Long Dark, we can begin to envision pathways forward that prioritize life, ecosystems, and connection.

The third type of action required for the emergence of the heart-centered revolution is a *shift in consciousness*—an opening to the embodied connection coursing through us, the larger world, and the ever-expanding universe around us. Feeling disconnected from the world and each other makes it a lot easier to exploit life and life-supporting systems. When we sense and nurture our connection with all that is, we become more compassionate toward ourselves and each other. We realize that we are not individuals contained in our corporeal body but in a dynamic relationship with the air, water, and larger world around us. When we see our connection to all that is, we open our circle of care and concern to species extinction, the climate catastrophe, and the whole of humankind. We refuse to look away from the injustices perpetuating separatism, alienation, and Othering. This shift of consciousness ushers in connectedness, love, and community as we break down the barriers that make us believe in our separateness.

In some activist and academic circles, caring for oneself and caring for the world are seen as separate endeavors. But Macy and Johnstone express that when we tend to our inner growth, we deepen our compassion for others, extending the reach of our work beyond just ourselves. We also renew the courage we need to face the predicament of our times. And when we neglect our inner worlds, we can end up emotionally immature, resentful, shut down, or burned out. Our survival responses take over, perpetuating our isolation, alienation, Othering, and harm.

How to Use This Book

Each chapter of this book will outline a step in the GGN's program, illustrated by personal stories from the lives of the authors. We offer snapshots of our lives, our struggles, and our hard-earned lessons. By sharing some of ourselves with you, we aim to illuminate that being openhearted is an ongoing choice of letting our heartbreak transform and rearrange us, increasing our capacity for connectedness. This ongoing choice is not easy, but it is the only way to be fully alive in the midst of ongoing crises. Some of our stories touch on potentially triggering topics, including child abuse, neglect, suicidal thinking and attempts, and self-harm. **This is your content warning for the autobiographical portions of this book.** You are welcome to read our stories or skip them depending on your individual needs.

EXERCISES AS PRACTICES

The process this book lays out cannot be an intellectual exercise alone. Your life will not be transformed by reading these pages but by integrating the suggestions and larger questions into your life and creating a community to engage with. You do not need more information; you need embodiment and reconnection. You need practice and community. This book offers a pathway—one among many—but you must take the path.

Sprinkled throughout the book are exercises designed to help you radically reconnect to yourself, others, and the more-than-human world. In doing so, we shift our consciousness: our inner work informs the outer work we all must do in the unraveling world. We teach these exercises in our GGN programs and we utilize them in our day-to-day lives.

Some exercises may ignite something in you. Add them to your toolbox and practice them. Some may not do anything for you. Feel free to leave those behind. There are a million practices out there waiting for you to come into relationship with them. Learn and incorporate new ones into your daily routine. These practices may help you become present in the current moment, in your body, and in your relationships.

If you are new to welcoming your whole range of feelings, some of the exercises may throw your nervous system into dysregulation. We highly recommend listening to your body, going at your own pace, working the steps in community, or engaging in the practices with a *healer*. Before practicing, call in your resources. Your resources are your set of tools, either internal or external, that provide a sense of safety or overall "okayness."[8] They are tools that are unique to you. We invite you to explore the things, places, people, ancestors, or sayings that might help ground you through your times of activation. We offer an exercise on pages 110–11 that will assist you in the process of identifying your resources if you are having trouble thinking of them right now. When implemented, these resources help you return to a presenced, calm state after being activated. They allow you to work through discomfort or challenging experiences.

Before each exercise, ask yourself, *Am I relatively safe, free of ongoing harm, so that I can dig into each practice? Am I well-resourced, with tools and strategies that help me ground myself if I experience activation?* If the answer is yes to both questions, it is time to practice. For each exercise, notice how you feel as you engage with it. If you become activated, you can back out of it and explore why you are activated. Use your resources to restore a sense of okayness and see if you want to return to it. Maybe an exercise is best practiced with a healer or in community. We are trusting you to notice and attend to your own emotional needs throughout.

Please modify the physical or breathing exercises we offer in a way that works for your body and your situation. These are general exercises that can be explored and altered to fit your needs. We suggest that if something causes pinching or pain, ease back from the practice. If something activates you and you are not well-resourced, do not practice it. You are responsible for noticing and caring for yourself physically and emotionally through these pages.

Exercises are not enough. Alone, they will not overcome the systemic oppression or solve the climate crisis. However, engaging in practices that aid in calming your nervous system allows you to form deeper connections with yourself, others, and the more-than-human world. Practices that ground you in the present moment help you notice what is around you, what is still here and worthy of admiration and celebration. The practices, done time and time again, will open you to being fully present in each moment as you look for opportunities to take the next best step. A practice is a relationship and an act of accountability to assert your aliveness.

As you engage with the practices in the book, we recommend recording your thoughts and reflections in a journal. Explore your feelings on the page or in conversation with a partner, friends, or in community. Return to the exercises daily, weekly, or as often as you are able. Again, we will repeat that the journey cannot be solely an intellectual exercise—it is up to you to utilize this material and integrate what works for you, loosen your grip on your worldview, and create communities of practice where you can be supported through the transition.

The exercises are an invitation to rootedness so that you can quiet your mind and attune to your deeper soul calling out to you. We are summoning you to radically alter your life: steady yourself, roll up your sleeves, do inner work, and reconnect at all levels. There is some real work to do—and we do not mean "work" as defined by the dominant paradigm. We mean building relationships and continuing to keep our hearts open even as the world feels like it's crumbling around us. Our practices teach us to trust ourselves, collaborate with one another, plant seeds, and share resources. Our practices set the intentions for the world we want to create, moment by moment.

RESOURCES AND KEY TERMS

We love connecting people with books and other media that might help inform your journey. At the back of the book, we direct you to a regularly updated webpage offering a variety of resources curated for further exploration of the subjects covered in each step. Resource suggestions include a list of books, articles, videos, podcasts, and songs. Additionally, we are covering some pretty intense subjects. As you read these pages, we encourage you to regularly pause and move your body. Aimee, who enjoys DJing, has crafted a playlist (see page 283) that accompanies the book—we encourage you to put it on and dance it out as you read through these pages. In an attempt to bring us together in a common vocabulary, we have also included a list of key terms, which can be found in the glossary.

AN OFFERING OF IDEAS

We see ourselves as curators, collators, and mash-up artists who study, organize, and combine a variety of ideas and theories into one place.

We are bridges creating links from one island of thought to another. We bring in some large concepts in every step, each of which could easily be, and probably already has been, a dissertation or book. Our goal is to introduce you to a series of topics in hopes that you will further investigate the subjects that stir something in you. Carry the conversation further with your own exploration and dialogues within your communities. This book, and our support groups, are by no means a comprehensive or complete solution to the many crises that confront us. We aim to ask more questions than to present answers, to serve as pieces of the collective dialogue, pushing ideas forward. As we say in Adult Children of Alcoholics meetings, "Take what you like and leave the rest." Our hope is that the ideas we have collected here will inspire you to move the dial further. Keep the conversation emerging. This is not a time of solution making; instead, it is a time of undoing, grieving, and repairing. The next steps are to follow this essential work.

Those who have worked the steps in our program say they've gained a sense of clarity and calm and found new ways to make meaningful connections. As one participant, Zeinab, said upon completing Step 10, "For the first time, I'm living life with my heart open." This is our offering to you.

The Gift of Now

We are grasping creatures. It's normal to want to hold on to the familiar, especially when confronted with uncertainty. However, grasping onto a culture that is exterminating species, people, and a habitable planet is not serving us. It's time to let go of our illusions of stability and a certain future, and come to terms with the darkness setting in around us. Darkness isn't bad; it simply represents the unknown. As our vision of the future fades, our other senses are enhanced. We don't need light to make meaning and connections. Our spiritual mentor and Rites of Passage guide, Kinde Nebeker, once suggested to us that we are supposed to *feel* our way through the Long Dark.

The proposal to feel is foreign to many of us. Emotions are often considered counterproductive, distracting, and too feminine, which is deemed inferior by our patriarchal society. This cultural narrative around our feelings keeps us disconnected, constricted, and acting small. The time for acting small is over. Each of us is being called to break through the cultural conditioning that holds us back and keeps us perpetuating harm.

The invitation of this book, and of our present time, is to live in the moment. Be alive—*right now* with *what is.* Not tomorrow. Not next week. We are not guaranteed that time anyway. Let go of your expectations of the future so that you can lead an openhearted life in each moment.

No matter what is happening in the world at large, you can and must make meaning now. These steps pave a path for you to live an empowered, meaningful life—by helping to calm your nervous system, open to the wisdom deep within yourself, and see the magic in each day. Most importantly, these steps encourage you to connect, connect, connect—even when your nervous system tells you to consider everyone and everything into a threat. The most important actions we can take in this time of transition are to connect more fiercely, to love harder, and to show up as our beautifully flawed selves.

We are often asked by journalists, group participants, and, sometimes, our own doubtful voices inside of us, "Is it too late?" However, it is never too late to get closer to our values, to cherish more about being alive in this time of great transition, and rediscover our relationality with this dynamic world. It is never too late to protect those we love, and we love people from a variety of cultures living all around our planet. We also love the flora and fauna that call Earth home. We love orangutans, blue whales, scalloped hammerhead sharks, arctic foxes, and the island forest frogs. We are awed by the coral reefs, giant redwood trees, hummingbirds, honeybees, flowers, and mycelium. The rich variety of colors, textures, shapes, sizes and sounds that compose the creative expression of life are still worth protecting. Even if we do not have a heart-centered revolution fast enough to prevent catastrophic biosphere collapse, these steps provide a pathway to minimize our harm, protect what we can, see the best in people, make meaning while we're alive, and constantly look for alternatives to solutions presented.

The steps in this book have the potential to make us better people: more-loving partners, more-attentive parents, less destructive in our lifestyles, and more connected to reality and other beings. These values do not fit smoothly into the dominant culture, and you will almost certainly encounter some people who cannot understand the journey you're undertaking. You will constantly be reminded of how inconvenient change is and how countercultural you are (though there are worse insults we can think of). Your mind may tell you that your identity is wrapped up in being a successful agent of this culture, that you'd rather have a set of operating rules than the anarchy of confusion. To counter this, we exercise patience and compassion with ourselves and extend them outward. We lean on our human community and the larger world for strength and courage. We practice the steps so that the moments between crises—the time not interrupted by struggle—are sweeter, sharper, and more defined. This is what keeps people alive, unified, and working toward a better future, even if we have no guarantee we'll get to said future. As many have said, life is about the journey not the destination.

There can be no mistake about it: uncertainty is all around us. We will work to create ways of being that nurture relationships with the living planet of which we are a part. These ways of being are born of intentionality and breathed by a reverence for life. Those of us who survive the Long Dark will be alive because of our webs of connectivity. We will rise from the ashes of an unraveling world with seeds in our pockets.

We Belong to Each Other

THOUGHTS ON COMMUNITY

> It is probable that the next Buddha will not take the form of an individual. The next Buddha may take the form of a community, a community practicing understanding and loving-kindness, a community practicing mindful living.
>
> —Thich Nhat Hanh

LaUra

"I can't believe Kara's been gone a week," I said to Sarah and Kellie in our school's media center during fifth period. "It all just feels like too much . . ." My friends nodded in agreement. Kara Spindler died on a Tuesday. She fought a long battle with leukemia, and it was too much for her weakened teenage body.

"Maybe we could go to the Spindlers' house after school," Sarah suggested. The three of us exchanged glances. Should we show up unannounced at our dead friend's house? Would we be interrupting her family's grieving process?

"Well, we could drive by and see if they're home," Kellie added. That was plan enough for the three of us.

After the bell rang, Sarah, Kellie, and I loaded into Sarah's car and drove past the village limits to Kara's home. By the time we reached the front door, our friend's parents, Dave and Lyn, were standing on their stoop. They embraced each of us, eagerly inviting us inside.

Lyn set glasses of water for each of us around the kitchen table and launched right in. With tears in her eyes and a shaky voice, she told us of Kara's last moments: a request for pickles, the permission from the nurse to let go, and a final hand squeeze. We were all in tears. Taking turns, each of us added our own stories about our last moments with Kara. There were even moments of laughter. By the end of the evening, each of us were feeling a little bit lighter and a lot more connected to our deceased loved one and to each other.

The next Tuesday, Sarah, Kellie, and I showed up again. This time, we were also joined by Kara's brothers. Once more, we shared stories, tears, and a little bit of laughter. We felt more connected and a bit lighter. The magic worked again. The Spindlers invited us back the following week and offered a meal. They even suggested we invite more of Kara's friends.

Week after week we showed up. Some weeks we had fifteen people and others only the core group of five of us. This tradition that we called "Tuesdays" became a weekly ritual where anyone mourning Kara could come, share a meal together, and offer stories. We didn't try to fix each other. We couldn't fix each other; there was no way to bring Kara back. Instead, we played games and connected. We asserted that joy and meaning existed even among the heaviness of death.

Months later, this gathering became a place to hold Kellie's family members after she was killed in a head-on collision. We continued to show up for each other, for the Spindlers, and now for Kellie's family. The container of the Spindler house on Tuesday nights was big enough to hold all the confusion, pain, and grief, as well as the celebration of life. Tuesdays lasted for years and moved each of us toward our own version of healing as we struggled with how to make sense of the magnitude of loss.

After plunging myself into a variety of healing modalities, one universal lesson became unavoidably clear: we need authentic community where we can be held in the entirety of our messiness. Human beings are deeply social creatures, dependent on one another for connection and survival. When we connect with open hearts, an alchemical process unfurls that is capable of transforming pain into meaning and isolation into connectedness.

Those of us carrying deep wounds must learn how to be with one another again. We are facing multifarious collective crises of which the climate emergency is the largest. The climate emergency is a crisis of connection; our disconnection from others and ourselves allows us to be complicit as the dominant culture pillages our ecosystems, species, populations of people, and a habitable biosphere. Surviving this time of planetary rearrangement requires reconnection. We will need to lean in and learn to trust and depend on each other to mitigate the worst impacts of the climate emergency, cocreate life-centered futures, and survive the tidal wave of social and ecological disorder headed our way.

Being socialized in the dominant culture is a lot like being raised in a dysfunctional family. We learn that we don't talk about the depths of our emotions or experiences, and we certainly shouldn't process our painful moments together. We are left alone to heal, keeping our pain private. But our grief is a universal connector—and a powerful one. Many of us live overwhelmed by loneliness and disconnection yet yearn to be in authentic relationship with others. I witnessed through Tuesdays, that by its very nature, healing is a collective endeavor, a community enterprise.

A healthy, vibrant, and nourishing community is dependent on a healthy, vibrant, and nourishing planet. For far too long, we in the dominant culture have designed our social systems without a deep understanding of our interdependence with the world outside of us.

Community is much more expansive than just a social practice. Our health and well-being are tied to the health and well-being of our life-supporting systems. If our air, water, soil, and ecosystems are degraded, we cannot expect to be well. We are in a dynamic relationship with the world around us, taking in and excreting molecules and energy. As we poison our planet, we poison ourselves.

Community has become a buzzword as of late. This is great news because it points to a profound stirring within so many of us to reconnect. We've forgotten how to be with one another. Practice groups can help in this remembering. GGN's 10-Step program offers an example of how we might be together in a microcosm that expands into the world. As we move toward creating strong local communities, connection and support become increasingly important to survive the Long Dark with meaning and joy.

Creating Community

> A good activist builds community.
>
> —Terry Tempest Williams

Creating community is an art, not a science. It's a cultivation, a practice, a remembering of the ancient ways of how to be with one another and how to be in relationship with the more-than-human world.

Connecting authentically requires us to do dedicated deconstruction work. First, we must identify and break down the walls that keep us disconnected from ourselves. This work asks us to notice and heal that which keeps us from being embodied, from feeling our sensations and processing our feelings. Next, we are called to notice the barriers that separate us from the world outside of our own skin. Finally, we must cocreate ways of being with each other that are not based in oppression, exploitation, or hyperindividualism but instead flow with belonging, reciprocity, and cooperation.

As we remember how to be in community with each other, we relearn that trust is built by strengthening relationships through reserving our judgment and practicing openness, vulnerability, and mutuality. In the words of the facilitator, author, and visionary adrienne maree brown, "Trust the people. If you trust the people, they become trustworthy."[1] This is just one of her nine "emergent strategy principles" that help lead us into the cocreation of more just and life-centered paradigms, and we explore these principles throughout the book. When we cultivate trust with those around us and in our community spaces, feelings of profound belonging follow. When we have a felt sense of belonging, collaboration increases, we open to new ideas, we feel more courageous to enact change, and we are more willing to support each other.

We, LaUra and Aimee, don't have all the answers about how to create a strong, trusting community based in belonging, reciprocity, and cooperation. But we are practiced in creating community groups that trust and inspire each other and continue to be informed by presencing, curiosity, and openness. We are creating as we go and looking beyond the dominant culture for lessons in how to reconnect.

If a future focused on equity and life seems out of reach, we can look to the models that have been experimenting with building communities that empower and care for those most marginalized. We can explore and learn from the Transition Town movement in the United Kingdom, Cooperation Jackson in Jackson, Mississippi, or Barcelona en Comú in Barcelona, Spain.[2]

Lessons learned from the ecological world can also guide us toward the cocreation of connected and resilient communities. Through biological sciences we have learned that the whole system is dynamic and made of innumerable parts and players, each contributing to the bigger picture. We see that increased genetic diversity, as well as diversity in populations within a community, strengthens ecosystems against pests and disturbances. Practicing mutualism, where multiple parties come together to meet a need and each party benefits, is another strategy we can carry over to our community-building spaces. Another ecological lesson we ought to apply to human communities is the ability to act with restraint, allowing energy and nutrients to cycle through systems. Energy and resource hoarding is not a behavior we see naturally occurring in the more-than-human world. Each of these strategies increases the health and resilience of the ecosystem. Because nature has been designing life and life-supporting systems for billions of years, we can emulate successful designs with humility to help us create healthy and resilient human systems. We see these efforts being implemented by the biomimicry movement and the permaculture community.[3] What other initiatives look to incorporate natural designs and techniques?

Many of these lessons are already known by cultures living outside the dominant paradigm. Most of the remaining intact ecosystems are protected by Indigenous land keepers and local communities.[4] These communities do not set land aside to keep as "pristine" but are in intimate relationship with it. For many Indigenous communities, humans are part of ecosystems not separate from them. They have shown us that we can live in right relationship with the more-than-human world while also protecting it.

As we cocreate new paradigms and build community, we ought to be regularly asking ourselves, *Who are the people and what are the models to learn from? What do ecologically aligned communities look like? How can we elevate the voices of Indigenous peoples who have known how to create community for thousands of years? Whose voices in community building have been left out of the conversation?* Let us start moving in the direction of those answers and keep these questions alive as we journey through these pages.

EXPLORING OUR BARRIERS

The following questions may help us identify how we have created barriers in our lives:

- Who and what do you perceive as separate from you? Why do you think you have created this distinction? What is it serving to keep these things/beings separate from you? What harm is it causing by keeping these things/being separate from you?
- What are some examples of times you have felt connected and supported?
- What are some obstacles that prevent you from receiving support?
- What attitudes and beliefs do you want to bring into the future world? What is one step you can take to help these attitudes and beliefs make it into new paradigms?

We Become What We Practice: Practice Groups

> Community is a choice. More precisely, community is an accumulation of choices made every day, a set of growing practices.
>
> —adrienne maree brown

Social and ecological crises demand change from the dominant systems. And change is hard, especially if we are trying to change our habits, norms, and worldviews. We need the support of a loving community to help us make transitions in our ways of being. But how do we get the support of a loving community if we are embedded in business as usual (BAU)? We suggest practice groups.

Sometimes we are so disconnected from ourselves and each other that we require some assistance in remembering how to be together in community. A practice group is a communal space that serves as practice grounds for remembering how to be with each other with our armor lowered, which then allows us to move toward cultivating new ways of being.

In practice groups, community serves as a place to heal individually and create new opportunities collectively. We can change our behaviors and overcome addictions with the help of community. Twelve-step groups like Alcoholics Anonymous and Adult Children of Alcoholics have proven their ability to create community and sustain behavioral change using a peer-to-peer support model. We are social creatures, cooperative animals. Behavior change for individuals is hard. Unseating our deeply ingrained patterns becomes a whole lot easier if (1) others around us are making similar changes and (2) we are supported and encouraged in our transformation.

Coming together with our community allows us to examine ourselves as individuals. We can reflect on our values and notice any barriers to connection that we are holding on to. We gather in community to remember how to feel our feelings, listen for our personal and collective truths, and establish healthy boundaries. This healing work brings clarity and calm so we can begin to deconstruct the dominant cultural narratives that live within us. Because this requires exceptional courage, each of us must voluntarily show up for it. No one can do this work for us. Each of us is responsible for our individual and collective transformation.

SUGGESTIONS FOR CREATING COMMUNITY PRACTICE GROUPS

Working within practice groups can sometimes generate conflict. That is why we need a set of agreements for how we will carry ourselves through the hard work of being together and cocreating new ways of being.

Here are some suggestions we use in our GGN spaces to help remind us how to be in community.

- Create community guidelines for how to be with each other

and a plan for what to do if the guidelines are breeched. Create space for accountability and repair as harm occurs.

- Practice full acceptance of what someone else is experiencing and expressing, even if you do not agree with it.
- Practice active listening. Listen not to respond but to hear, to connect, to empathize.
- Be with someone's experience without attempting to fix them or their situation, even if it makes you uncomfortable. Suspend the urge to give advice.
- Allow for the expression of a full range of feelings, each other's and your own.
- Remain open and curious. Notice when you are no longer in that state and practice grounding techniques to get you back there.
- Practice courage and vulnerability.
- Approach your community nonjudgmentally.
- Empower each other and yourself to trust your own inner guidance, evolutionary impulse, and intuition.
- Pursue individual and collective actions that are in service of connection, growth, and healing.

Brave, Emergent Spaces

> Perhaps the secret of living well is not in having all the answers but in pursuing unanswerable questions in good company.
>
> —Rachel Naomi Remen

We need spaces where we can be brought out of our comfort zones and to the edges of our understanding and perception. We do not change when we are comfortable. We need the motivation of discomfort to transform ourselves and our communities so that we might try on new ways of being. Fortunately, these times provide endless opportunities for us to dissolve and reconfigure ideas, norms, and habits. If we remain open and curious, we can notice the edges of our comfort and begin exploring them with the intention to stretch them. With one foot in the new and one in the old, our edges are where we can courageously challenge our discomfort to glimpse new ways of being.

We do not have to go to these edges alone; we should not. This work is what practice groups are for! In a group setting, we can bravely explore our edges alongside others asking similar questions.[5] Cocreating just and life-centered futures requires risk and embracing uncertainty. It will not always feel safe, comfortable, or familiar because we need space to be with, explore, and deconstruct portions of ourselves and our collectives. Navigating our edges requires regular introspection and inner work. Many of us equate comfort with safety, and any experience of discomfort may lead to experiencing a trigger or our survival responses. If we go too far and do not have proper practices in place to support our healing, our defense mechanisms might become activated, shutting down the whole process. Moreover, if our edges were determined by trauma responses or other unhealed wounds, it might be best to stretch them with the help of a healer, outside of the practice groups.

As we start learning how to be with discomfort, this process can be helped by establishing a series of regular practices that help us ground and regulate our nervous system. It may take additional work, energy, and healing to be with discomfort in a generative way. Discernment, self-awareness, and trust can provide key information. The goal is to traverse our edges with a sense of calm and clarity. When we do this in community, we cultivate spaces that are fertile for the emergence of new ideas and perspectives. This is the start of creating new ways of being.

~

We have run several dozen 10-Step programs over the years and despite the consistency of the steps, each program is different. Where we end up as a community depends on who is in the room, their quality of showing up, and their ability to exercise courage to express new ideas. Together we expand our realm of what is possible. Another emergent strategy principle instructs, "There is a conversation in the room that only these people at this moment can have. Find it."[6] With a willingness of the unique people in the room to fully show up, we can arrive at a place we could not have imagined just moments before.

Communities are like the banks of a river. Each community member is a molecule of water. By connecting to other molecules, we exhibit bravery and openness. Connected, the molecules create a body of water that flows through those banks. We become freer to meander and explore. Over time, the water shifts the riverbanks and erodes river rocks. The river changes shape. Our communities of practice, too, can transform over time.

The growth edge is a precarious zone, capable of helping us formulate fresh ideas and birth new ways of being. We must be willing to be together in times of grief, anger, and despair while enduring the tension. We must be willing to live the questions together instead of rushing to solution making just for the sake of satiating the discomfort. We must learn to care for one another. We must feel our way through, together.

Community for the Long Dark

> A single twig breaks, but the bundle of twigs is strong.
>
> —Chief Tecumseh, a Shawnee warrior-chief

Change is coming, whether we like it or not. If we are not doing the work to regulate our nervous systems and come together, the overwhelm of the ongoing disruptions will fragment and isolate us from one another. We need each other to cocreate new ways of being and to survive the Long Dark.

We are in a tough spot. We find ourselves desperately looking to politicians worldwide to lead us through this catastrophe. Yet vested interests from multinational corporations maintain a stranglehold on top-down change. When we cannot look to them for strong leadership and we feel too small as individuals to effect meaningful change, what is left? This is where community-level action is a catalyst for creating equitable, life-centered ways of being.

MENTAL HEALTH IN THE LONG DARK

Those of us alive in this time are experiencing a myriad of overlapping factors that implicate our individual and collective mental health. In the past few years, humans have lived through several global, historic events with little guidance on how to deal with the acute, ongoing traumas, including the massive losses of life due to the COVID-19 pandemic, ongoing systemic social injustices resulting in mass murder and state violence, the weakening of democracy, and the varied impacts of the climate emergency. The state of the world, as it is now, has already created a mental health crisis. One 2021 study found that out of ten thousand young people surveyed between the ages of sixteen and twenty-five and across ten countries, 75 percent felt that "the future is frightening."[7]

In our GGN spaces, we regularly sit with folks overwhelmed by feelings of dread, despair, grief, hopelessness, and helplessness. And as the Earth warms, bringing with it unimaginable shocks, we will see an increase in psychological issues. In 2021, ecoAmerica and the American Psychological Association teamed up to issue the report *Mental Health and Our Changing Climate: Impacts, Inequities, Responses*. The authors warn of the looming mental health crisis due to the increased impacts of the climate crisis:

> Longer term climate change can cause equally significant mental health impacts. Heat can fuel mood and anxiety disorders, schizophrenia, vascular dementia, use of emergency mental health services, suicide, interpersonal aggression, and violence. Drought can lead to stress, anxiety, depression, uncertainty,

> shame, humiliation, and suicide, particularly amongst farmers. Air pollution has been linked to increased anxiety and use of mental health services, lower happiness and life satisfaction, and other negative well-being impacts. Changes in the local environment can cause grief, emotional pain, disorientation, and poor work performance as well as harm interpersonal relationships and self-esteem. Displacement can cause a range of negative mental health impacts due to loss of place, community, and livelihoods. The loss of personal identity, autonomy, control, and culture can lead to mental distress, insecurity, diminished self-worth, sadness, anxiety, depression, anger, and weakened social and community cohesion. A warming climate can also lead to aggravated interpersonal aggression (such as domestic violence, assault, and rape) and interpersonal violence (murder). Heightened anxiety and uncertainty likewise negatively impact social relationships and attitudes toward other people. Migration and competition for scarce resources can lead to intergroup hostility, aggression, violence (political conflict, war), and even terrorism.[8]

While the worst consequences of the climate emergency lie ahead of us, the effect of multiple crises on our mental health is becoming increasingly apparent. Like the climate crisis overall, mental health impacts will be disproportionally borne by low-income, marginalized communities, youth, and women.

The ongoing biological and social disruptions will exacerbate individuals' mental illness. Folks like Aimee and me, with preexisting mental illness, are susceptible to our conditions worsening as we plunge deeper into the Long Dark. We need to practice methods for metabolizing the ongoing and increasing levels of distress, so we can continue to be open to connectivity, our imagination, and critical thinking.

Aimee and I have both been transformed by working with good therapists throughout our lives. We know the value and healing that can come from working in a therapeutic relationship. However, traditional modes of therapy are already insufficient to offer the level of help demanded by the ongoing biological and social disruptions. Currently, here in the United States, there are months-long waiting lists to see therapists and psychiatrists, access and affordability issues, a lack of clinicians who understand the predicament, and an overarching stigma against receiving mental health care. Furthermore, traditional modes of mental health care often individualize deeply collective issues, failing to incorporate the social context that causes much of the distress in our modern world. Currently, mental health care is palliative care and not actually providing healing for the social, political, and biological ailments that contribute to the mental health epidemic. Especially in the United States, clinicians are helping to pull some people from a rushing river, yet many fail to address the oppression, exploitation, and disconnection that has us falling into the water to begin with.

While writing this book, I was wait-listed for three months in the small Nebraska town where I was living before I was able to connect with a therapist who told me to quit my work with GGN because the world's problems are too big and there is nothing I can do about them. Although this is an extreme example, I have worked with a number of therapists in the past fifteen years who have minimized my concern over, and pain for, the destruction of our planet. GGN was created because I needed support to face the painful truths of this time: life is being decimated by the dominant culture, and the most vulnerable among us are suffering the worst.

Traditional models of mental health care need to head upstream as many continue falling into the river. In an interview, Susanne Moser, a climate change adaptation professional, explained, "Therapists, counselors, the mental health community in general, is very late to the game. . . . They're 20 years behind—at least."[9] To combat the mental health epidemic that is growing exponentially, we cannot use the same models we have been relying on for the past hundred years. There is great work being done by climate-aware therapists who are seeking to expand their practices and methodologies to help meet the demand.[10] There has also been movement to decolonize the traditional biomedical therapeutic model, expanding mental health care out of the Eurocentric models, making it more inclusive, and enabling it to serve a greater number of people.

As the mental health care model gets a makeover, we are strongly advocating for bringing practice groups to life in our localities. Because we know the mental health crisis is growing, and the current model has been outpaced by the climate emergency, we must cultivate strong communities capable of helping in ways we cannot even imagine right now. We—especially those of us in the dominant culture—are all being forced to grow, expand, and change the ways we have been doing things. For crises of disconnection and separation, like social justice issues and the climate emergency, we need a collective healing approach. Collective problems require a collective rewiring. We sense the absolute need to come together, learn how to work through instances of harm, and navigate the Long Dark in deep, authentic connection.

WE CANNOT GO IT ALONE

Community offers support, inspiration, and belonging to help us endure this time of rearranging and disorientation. As the world around us fragments and rearranges, community will help us take root. We can find grounding in our relationships instead of certainty and learn to find safety among one another. We are headed for disruption that we cannot even fathom. Since the pathways ahead have not yet been traveled, group consciousness and shared wisdom can add insight into how to navigate forward. We cannot survive in a tumultuous world without each other.

We need idealistic, pragmatic, conscious communities to grow and change. As the climate emergency eats up habitable land, making our living space more condensed, we will see an influx of people displaced by the direct impacts of the crisis (heat waves, storms, fires, flooding) as well as secondary impacts (food shortages, water scarcity, and resource wars). Statistics on climate refugees are difficult to quantify—how do we measure migration from direct and secondary impacts that are often coupled with political instability? Nevertheless, we know migrations have already begun. We have seen this throughout Central America and in Syria, in particular.

Everyone deserves safety, dignity, and belonging, including people who have had their homes ripped out from under them. These are basic human rights. We must extend our rigid boundaries of ownership and privatization and expand our notion of inclusion. How can we welcome climate refugees into our communities? What will happen when *we* are the refugees? Though the impacts of the climate emergency are borne disproportionally by marginalized communities, fires, floods, food shortages, extreme weather events, and the like will affect us all.

Ahmed Ali
@MrAhmednurAli

Friendly reminder, that we are all closer to being climate refugees than billionaires.

6:44 PM · 7/16/22 · Twitter Web App

A viral tweet from the health policy researcher Ahmed Ali

Starting right now, we can begin practicing new ways of being together, planting and nurturing the seeds that will take root for generations to come. As we endure the Long Dark, our communities will serve as inspiration, belonging, and support. We offer our 10 Steps as a map for cocreating heart-centered communities.

Coming together in community is an integral part of each of these steps. Community can hold and help heal deep wounds. When our home is flooded or we cannot afford food, it is our community members who come to our aid. And crucial for our moment in history, community has the power to withdraw from the dominant culture and cocreate new ways of being that prioritize care and connection. As a single person, we are limited. As a community, we are strong and supported. We are social creatures, and community is required for long-term, drastic transformation. No one person has all the answers to the many problems we face. No lone hero is coming to save us. It is up to each of us to be courageous and work together to create pathways of meaning, connection, and joy throughout the Long Dark.

With this in mind, we urge you to seek connection with others as you work the 10 Steps in this book, whether with a healer, a loved one, a reading circle, or a community of practice. Discuss these questions together: What do the ideas on these pages make you feel? What is stirring in you? What are your next steps that contribute to the healing of the world?

We invite you to keep community, inclusion, and expansion in mind as you do this work.

CREATING COMMUNITY IS CRITICAL

- Cocreating equitable, life-centered futures requires collaborative ideation, inspiration, and collective change.
- Community provides rooting and groundedness and the encouragement to do hard work.
- Community provides mutual aid after an extreme weather disaster or when community members cannot afford to buy food or pay bills.
- Community provides a consistent source of belonging, connection, and safety.

Step 1

ACCEPT THE SEVERITY OF THE PREDICAMENT

And once you are awake, you shall remain awake eternally.

—Friedrich Nietzsche, *Thus Spoke Zarathustra: A Book for All and None*

Aimee

This step is the hardest. I still struggle with it. Dozens of times now I have worked GGN's 10 Steps to Resilience & Empowerment in a Chaotic Climate program, and I always come back to Step 1 with deeper grief. Wait—*deeper* grief? Isn't climate grief the problem and our 10 Steps the solution? I get it. You're looking for that magic prescription to make everything hurt less, and I'm with you. It's hard to be a critically thinking, emotionally intelligent person in today's world. But here's the thing: collective grief isn't a problem to eradicate. Nor is rage, fear, or any other feeling you are experiencing in response to the state of the world. These are totally normal human responses to a hurting planet. The uncomfortable feelings that keep us up late at night over our worldwide predicament show that we feel, care, and are open to what is happening. These feelings aren't comfortable, but opening to them is the most human response you can have, and this all is a vital step toward transformation. Step 1 is meant to normalize our emotional reactions to the state of the world and orient us to be a catalyst for profound change.

I was much slower than LaUra to accept the severity of the predicament. We were undergraduates at Central Michigan University when LaUra—my best friend, now wife—started talking about the climate crisis as an existential threat not looming in our future but happening now, in our lifetime. At the time, I was active in our college's chapter of Amnesty International, protesting torture at Guantanamo Bay and registering students to vote. The more I learned about human rights violations around the world and within my own country, and the more I saw what I thought was our collective apathy toward this vast suffering, the more depressed I got.

It is an oversimplification to blame my first mental breakdown at age twenty-one on my activism. Like so many of us alive today, I come from a long line of addiction and mental illness. I was also a pageant kid, an overachiever, someone who didn't know how to take breaks, and who didn't know that failure is always an option, not a death sentence. I was raised by hardworking, middle-class parents who, with the best of intentions, taught me that the way to deal with pain was to ignore it. Repress uncomfortable feelings, suck it up, and keep working toward that American Dream. Paradoxically, it was that repressive mindset that both created my depression and helped me cope with it, until eventually, this coping tool stopped working.

After a handful of hospitalizations in a matter of months, the psychiatric doctors advised me to take a break from my activism to focus on my health. They assured me the world's problems would be there when I was better—only I couldn't seem to get well enough to return to it. My experience of depression was chronic and often treatment resistant.

While I took a step back from activism, LaUra dove further in. She often spoke about the severity of climate change, and I would nod, unable to fully take it in. If what she said was true, why were politicians and leaders denying its existence? Why were we not declaring an emergency? Surely the media wouldn't let such destruction happen. It was the journalists' job to maintain accountability, right? The idea of an injustice so deep, a future so uncertain, was too much for me to accept, especially in my already vulnerable emotional state.

My faith in the media vanished after LaUra moved to Louisiana to work for AmeriCorps directly following the British Petroleum (BP) oil spill in 2010. While I finished my undergraduate degree, LaUra witnessed firsthand the injustice perpetrated on frontline communities by the fossil-fuel corporation. In addition to the hundreds of millions of gallons of crude oil spilled into the Gulf of Mexico, BP pumped nearly two million gallons of a harsh chemical dispersant, Corexit, into the water to dissolve the oil. The combination of dispersant and oil proved to be a toxic duo, an impact BP hid from the media, claiming that the dispersant was as benign as dish soap. For months after the spill, there was a regional moratorium on fishing, shrimping, and pumping oil. Because of the harmful combination of oil and dispersant, there were massive die-offs in the Gulf, and surviving sea life was deformed. Researchers pulled out shrimp with no eyes and crabs with no claws.

Meanwhile, the local community was suffering the impacts of job loss and sickness. At a town hall LaUra attended in New Orleans, several people spoke of the health impacts they suffered after being exposed to the waters in the Gulf, ranging from bleeding from the eyes, ears, and anus to memory loss and other lingering physical effects. A young local man who had gone swimming after the spill and Corexit application blamed his seizures and need for a wheelchair on the chemicals in the water. "Once doctors heard I swam in the Gulf, they refused to treat me. The problems got worse eventually landing me in a wheelchair," he explained. His testimony was cut short by a seizure he suffered on stage.

One day, as I was watching TV with my parents, I was shocked to see a news anchor claim that the oil spill was largely cleaned up. I told my parents about Corexit, eyeless shrimp, and the man in the wheelchair who had bled from the ears and had a seizure in front of LaUra.

"No offense to LaUra, but that sounds like an exaggeration," said my dad. "They just said it is already cleared up."

No matter what I told them, my parents would not or could not accept the truth. As I think back on their reaction, I already knew the problem was bigger than the climate emergency alone, especially because I had been a social justice advocate already. This was my first exposure to the complicated interplay of climate, the outright lies of big corporations, and social and economic injustice. To have the reality I saw batted away by my family made me question my own reality. It made me question LaUra's. It is no wonder I struggled for so many years to accept the severity of the predicament. Like so many of us, I had learned from the culture around me how to minimize or deny truths that made me uncomfortable—how to look at problems in a piecemeal fashion and downplay their severity, even as the severity was right in front of me.

AN INVITATION TO ARRIVE HERE WITH YOUR FULL SELF

Pause. Take a breath with a deep inhale and a long exhale (a breath where the exhale is longer than the inhale). Turn your attention inward. Become more present in your body. Scan your body from your toes up to your head for any feelings or sensations. Do you feel tension, lightness, ease, a certain temperature in your body? Where in your body are sensations present? It is OK if you do not yet feel sensations in your body. With an intentional practice, you can begin to notice the sensations and feelings coursing throughout your body. As you check in more frequently, they will become more apparent. Your body communicates not through words but sensations. It is a source of wisdom that many of us have forgotten to include. Our body provides an honest report of how we are.

Turn your attention outside of yourself. What is directly around you? How is the air interacting with your skin? What scents are around you? What colors do you notice? Can you hear the sounds near you? Are there other beings near you?

Much of this work is about slowing down enough to notice your inner and outer worlds. It is about bringing your full self to the present moment and being with what is there. An embodied awareness can help you arrive with your entirety. We invite you to arrive to these pages with your entire self, each time you open the book. A deeper invitation exists to fully inhabit each moment of your life with as much presence as possible. This type of presence will help inform new ways of being. It is important to start practicing right now and as often as possible.

The Power-Over Structure

> The truth does not change according to our ability to stomach it.
>
> —Flannery O'Connor

Understanding power dynamics can play a key role in grasping the moving parts of the Great Unraveling. The increasing division between those who have power, wealth, and resources and those who do not has been perpetuated through our political, cultural, social, corporate, and economic structures through a system of domination best described as *power over*. In *The Politics of Trauma*, Staci K. Haines explains that this system concentrates "safety, belonging, dignity, decision-making, and resources within a few elite, and particular nation-states. This is done by taking from and exploiting others and the natural world. Those who are harmed and made poor are blamed in border social narratives."[1] In other words, our current power structure allows some to thrive because others are oppressed, and those who are oppressed are blamed for their own oppression.

Capitalism is the epitome of a power-over structure. It is characterized by the small number of rich and powerful elites (transnational corporations, the CEOs who run them, and the politicians who back them up and get rich in the process) and those who are marginalized (many people of the global majority, BIPOC communities in the Western world, women, etc.). Haines points out that cultural narratives that reinforce societal divisions do this by normalizing war, inequity, patriotism, and nationalism and are supported by the media, governments, and religion. Meanwhile, information showing us different perspectives is suppressed and minimized.

Because this power structure pervades all social systems in the dominant paradigm, it is aggressively utilized to make everyday people question our own realities. There are many tools employed by the power-over structure (e.g., violence, coercion, fear tactics, and misinformation campaigns), and we will highlight a few below.

COLLECTIVE GASLIGHTING: DENIAL BY DESIGN

The denial about the unfolding climate emergency is primarily in the United States and is orchestrated by a handful of CEOs, lobbyists, politicians, and corporations who see climate change mitigation as a threat to their profits. Doubt about the climate crisis was intentionally sown by companies like ExxonMobil, Texaco, and Shell, who paid billions of dollars to fund disinformation campaigns. In 2021, a bombshell paper came out and videos were leaked of oil lobbyists admitting they knew the urgency of the climate crisis.[2] They had data showing that continued fossil-fuel use would scorch our planet. Moreover, they ramped up a campaign to cast enough doubt to prevent phasing out of fossil fuels. They intend to profit from this way of being as long as they can.

We call this misinformation tactic *collective gaslighting*. Collective gaslighting is a tool of a power-over structure that operates in the dominant culture. The term *gaslight* comes from the name of a 1944 play, later turned into a film of the same name, in which a newlywed named Paula is tricked by her manipulative husband, Gregory, into believing that she is insane. His tactic: dimming and brightening the gaslights around the house, insisting that she is imagining it. It does not take long for Paula to begin to doubt herself and her own sanity.

Today, we are watching this script play out right before our eyes as politicians and corporations tell clever lie after clever lie. Yet collective gaslighting goes beyond lying about facts. It includes a key tactic shared by domestic gaslighters: victim blaming. Corporations and politicians use this technique to shift the focus onto individuals who have little influence over big systemic issues like the climate emergency, for example.

A classic example of corporations shifting blame onto individuals is the idea of the individual carbon footprint. Instead of holding big corporate polluters accountable, we are encouraged to track how much carbon we as individuals and family units consume. We are encouraged to change to LED light bulbs, drive and fly less, and turn our thermostat down a degree or five. In a powerful exposé, the journalist Mark Kaufman sheds some light on the public relations success that BP employed to distract individuals from holding big polluting corporations responsible for their greenhouse gas emissions (or their oil spill).[3]

BP created an individual carbon calculator in 2004, and it has been a popular tool ever since. While many of us are busy guilting and shaming ourselves and each other, multibillion-dollar industries continue tearing up our ecosystems, recording unfathomable profits, hiding their scientific studies that uncover the harm of the climate crisis, and making plans to drill in the Arctic as it thaws. We have been duped by clever propaganda campaigns into focusing on the micro level instead of demanding all fossil-fuel corporations cut emissions and move toward cleaner energy.

One critical tactic that sets collective gaslighters apart from domestic abusers is the power to create foot soldiers of denial. That is, collective gaslighters can turn ordinary individuals into skeptics of climate change, for example, who then sow the seeds of doubt for them for free, further normalizing the climate debate. By effectively fooling people into believing that climate change is not real or is not a serious, life-threatening issue, power-over structures fuel everyday gaslighting scenarios. Anyone who has fought with a family member about climate change over Thanksgiving dinner or who avoids interacting with their neighbor because they tell you all about what they heard from a Fox News pundit and it feels like you are living in two different realities has had the very painful experience of collective gaslighting.

Collective gaslighting is not just used to deny the reality of the climate emergency; we see it being employed on the social justice front too. Socially, BIPOC and other marginalized communities find themselves being blamed for their own oppression and then expected to somehow pull themselves up by their bootstraps despite hundreds of years of oppression and government regulation working against them. Our entire social structure is designed to divide and blame, continuing the cycle of denial in a way that is deeply personal and wounding.

These tactics of self-doubt and division, when utilized over and over again, cause us to fragment collectively and individually. We doubt ourselves and our intuition. We doubt each other. To survive in this destructive climate, it becomes safer to cast away our souls and other vulnerable parts of us. This disconnection, this soul loss and fragmentation, is at the root of our predicament.

BREAKING THE TRANCE

Even the most critical thinkers among us are susceptible to collective gaslighting and other power-over tools because of the *consensus trance*. Coined by the psychologist Charles Tart, the term *consensus trance* refers to the shared assumptions and agreements that bind communities under the spell of groupthink. Consensus trance has been described as a waking trance in which people believe what they have been indoctrinated to think, and as the author and teacher Terry Patten explains, "This mass hallucination is powerful, often impervious to our own direct experience and critical thinking."[4]

Gaslighting plays a huge role in keeping us in the consensus trance. In following an official narrative that has been crafted for us, we overlook our current reality to fit in with those around us. Tart calls this *consensus reality*. We conform with our in-group's collective agreement about reality and throw out, avoid, or deny unpalatable truths—like those pertaining to the climate emergency or systemic social injustice, for example. According to Tart, consensus trance is more than just a willingness to play along with others; it is an entire state of consciousness, one that is incredibly difficult to see beyond. Think of it like the simulated reality in the movie *The Matrix*. When you are inside the Matrix, you do not even know that you are operating in a consensus reality. You wake up from the trance and the larger reality only when you are unplugged.

We are assuming that if you have picked up this book, you have unplugged from the Matrix and are searching for other ways to be. As your consensus trance breaks, reflecting glitches in the consensus reality, you may wonder what is real; you may question your own sanity. If that's true, then you are in good company. *You are not alone.* So many of us are waking up, all across the globe. Those of us who are only waking up now have a lot of work to do, and we want to connect. We want to dream new ways of being that lay the foundations for life-centered paradigms together.

Many participants in our GGN support groups often come to their first meeting and confess that they have been made to feel crazy for being so upset about the state of the world. Some common refrains from our Step 1 meetings include the following:

> "I cannot talk with my friends or family members about the climate crisis or social justice issues. They all think I'm blowing

things out of proportion and that I should not worry about it so much."

"I feel like the world is falling apart and I'm the only person who notices or cares. Am I just overthinking how bad things are? It makes me feel insane."

"I think about these things all the time, and I cannot stop talking about them. I do not want to overwhelm my family and friends with my constant rumination. It leaves me feeling so alone and isolated."

The simple acknowledgment that, yes, the predicament we face in the world today really is as bad as we feel it is, is healing in itself. It removes the isolation many of us feel as we break through the consensus trance. By the time our participants leave their first meeting, they often feel relieved. They are not crazy after all. Chronic anxiety, depression, isolation, and psychological trauma are all long-term effects of gaslighting.[5] Our participants' grief and loneliness are not imagined—they were created by cruel design to support the consensus reality.

We are all the victims of collective gaslighting. But here is the catch: in many ways, we are the benefactors as well. That is one quality of collective gaslighting that sets it apart from domestic abuse: unlike interpersonal gaslighting, collective gaslighting benefits both the abusers and the victims. While corporations and politicians lie to continue BAU, those of us in the most privileged sectors of the dominant culture get to go on living our lives as we have been for decades, until we cannot anymore. It is no surprise that those of us deeply embedded in the consensus trance hold the most privilege. We are encouraged by sentiments within the dominant paradigm to own multiple cars per household, live in the suburbs and commute to work, order packages from Amazon, or snack on factory-farmed, genetically modified, pesticide-laden foods that ravage our ecosystems and our health. It is how our systems are built. Our participation in BAU, our perpetuation of our convenient lifestyles, directly contributes to the destruction of life and life-supporting systems. This painful reality makes us want to believe the lie, to continue the consensus reality. We may feel entitled to our comfort and luxuries and feel despair when they are threatened.

A NOTE OF CAUTION: ANGER AND RAGE

Feelings and emotions are complex. There is ongoing research looking to parse out different emotions, especially climate emotions.[6] While we honor the complexity and differences of emotional experiences, we also want to offer a note of caution. As we work our way out of the consensus trance, and we begin to understand the multiplicity of harms caused by the consensus reality and power-over structure, anger and rage are normal, natural reactions. We may feel anger and rage toward politicians, corporations, and the elites who are actively causing harm, who know about the consequences of their actions and still prioritize profit over life.

While anger and rage can be responses to boundary breaches and injustices, these feelings run hot and require an exceptional amount of energy to sustain. To dissipate the heat and quell these often-uncomfortable emotions, we might rush into action and cause harm. Or we may hold on to these searing feelings without processing them, causing us to burn out.

We can transform our raw, unprocessed anger and rage into motivation and sacred rage. This type of energy provides fuel in our fight to overturn long-held systems of oppression and exploitation. Transforming our raw heat into a righteous or sacred experience starts with attending to these feelings with love. We recommend sitting with the anger or rage (we offer an exercise on pages 67–68 that helps us learn to be with our feelings) to see if there might be other feelings underneath it. We have experienced that beneath our anger and rage, there is often immense grief, despair, betrayal, and even guilt asking to be felt and processed. We can stay with the heat of our discomfort to ask the anger or rage what it needs and what it has to teach us about our current situation. As we stay with the feelings, we create space to discern which reactions are in line with our personal and collective values. What course of action exercises compassion and can be conducted with integrity? As we act in a centered, more balanced way, we act on behalf of life, which opens us to sacred and righteous actions.

The Unsolvable, Complex Predicament

> The world hasn't ended, but the world as we know it has—even if we don't quite know it yet.
>
> —Bill McKibben, *Eaarth: Making a Life on a Tough New Planet*

Step 1 in our program—Accept the Severity of the Predicament—uses the word *predicament* instead of *problem*. There are many thinkers using the term *predicament*. We are not the first. The most concise distinction between the two words comes from Vanessa Machado de Oliveira in *Hospicing Modernity: Facing Humanity's Wrongs and the Implications for Social Activism*. Machado de Oliveira offers that *problems* are "things that can actually or potentially be fixed," and *predicaments* are "things that must constantly be dealt with, won't be solved, and won't go away."[7] She continues, "There is a difference between something *complicated* that can be sorted with careful planning or engineering (e.g., a long car trip with toddlers) and something *complex* that is moving, multidimensional, and largely unruly, unmanageable, and unpredictable (e.g., raising children)."[8]

We at GGN are widely known for helping people with ecoanxiety and climate grief, but it is never just about climate or ecological destruction. As we dig deeper and tug at one thread after another, we notice the many ways our current systems have failed us and our planet. It is never one singular issue, and to keep it focused on a single issue, a complicated problem, is to perpetuate the same piecemeal solution making that got us into this complex predicament to begin with.

Single-use plastics are a complicated problem. The climate crisis combined with our collapsing ecosystems and the onslaught of social issues spanning the entire globe, plus the campaigns designed to deny them—*this* is a complex predicament. The increasing complexity of interactions fueling biological and social breakdown means that we are in a time of rearranging, a time of descent and reprioritization. In a 2020 online teach-in called "The Pandemic Is a Portal," the writer and activist Arundhati Roy taught that we are here, in this moment of overlapping and complex issues, because a million decisions have been made, like threads that have been interwoven and interlinked to make a tapestry. The work of this time is to undo the tapestry one stitch at a time.[9]

Perhaps the climate emergency was simply a problem three and a half decades ago, in 1988, when NASA scientist James Hansen testified in front of the US Senate that climate change was a presently unfolding crisis that must be seriously addressed. But the consensus reality has ensured that we have pushed the climate emergency out of sight decade after decade as the dominant human civilization continues to "advance" our technologies, carving up ecosystems and increasing our greenhouse gas emissions. The impacts can no longer be pushed out of sight; they are closing in on us from every direction—melting polar ice, droughts, heat waves, extreme storms, wildfires, pandemics. There is much talk among scientists and politicians about the amount of warming our planet can withstand without triggering sensitive climate tipping points.[10] We hear a lot about keeping global warming within 1.5°C, but the truth is that we have already violated a number of planetary boundaries through human activity, making Earth's systems much more vulnerable. Even more fear inducing is that we do not know with exact certainty the level of warming that will trigger those climate tipping points whose cascading impacts will be abrupt, exponential, and irreversible. Indeed, we have peer-reviewed research and sophisticated models but still, many complex uncertainties remain. And every day, we get closer to the abrupt, exponential, and irreversible changes. While it seems that we are hurtling ourselves toward those tipping points, every tenth of a degree of warming matters to preserve life on Earth.

To limit warming and survive the impacts of the climate emergency, "rapid, far-reaching and unprecedented changes in all aspects of society" are required.[11] These words are quoted from the Intergovernmental Panel on Climate Change (IPCC) website, calling for a *reordering of every facet of the dominant paradigm*. How we cultivate and distribute food and goods, how we heat our buildings, ongoing wars, the ways that we interact with one another on a day-to-day basis—everything must change. This is a tremendous opportunity to reimagine our systems, our connections, our ways of being with each other and the Earth as a whole. But we cannot get there on fossil fuels and exploitation, which means changing nearly every way we in the dominant paradigm operate.

This is the truth of the predicament: whether we like it or not, everything will change. This is our entry into the Long Dark, the descent down the mountain of infinite growth and transition away from BAU. It is a time to rearrange our priorities. And following the instruction from Bayo Akomolafe—teacher, psychologist, and chief curator of the Emergence Network—we must get lost, become unmoored, question our questions, and find comfort amid the discomfort.[12] In Step 2, we explore how to be with uncertainty, but for now it is important to know that this perspective is not throwing in the towel. We are not advocating for the "do nothing, our planet is fucked anyway" route. Yet a transformation to truly just and life-centered paradigms may seem impossible to envision right now, especially when we have biological tipping points threatening the stability of our biosphere and our social structures are fracturing by the day.

Facing an unsolvable predicament is difficult work. Our pathways forward are holistic ones that include our bodies, minds, and souls and reaching out and forming strong communities. As we radically reconnect, we realign with the collective and uncover wisdom that has been pushed aside by keeping up with the frenzy of the dominant culture. What we are not advocating for is the rush to Band-Aid solutions that perpetuate the deep systemic problems that created the climate emergency to begin with. We are calling for difficult but meaningful emotional, psychological, and spiritual work that helps reweave the social fabric as we break free of the consensus reality that prevents us from envisioning the pathways of the future.

This journey will have you asking yourself, *Is there hope? Is hope necessary? What is the point of this hard, uncomfortable work? What power do I have, as one person, to fight multibillion-dollar corporations and propaganda machines? Why fight the good fight if it is all for naught? Why are we here? Why am I even alive? How can I identify the next step when everything seems dark?*

If some of these questions resonate with you, you are on the right path.

EXERCISE: ESTABLISH A JOURNALING PRACTICE

We recommend that you start a daily journaling practice to begin exploring your emotions and patterns of thinking. Stream-of-consciousness-style journaling is an excellent way to observe your thoughts, get to know your feelings, and move toward accepting your whole range of experiences. We strongly believe that, while there are painful and heavy feelings, we don't need to perceive these as negative. Time spent journaling does not have to be long. Sentences do not have to be perfect. And it is OK if you miss some days. But writing things down gets them out of our thoughts and allows for some perspective. Here are a few journaling questions to get you started:

- What is happening in your internal world when you notice that you feel overwhelmed about the climate crisis and the larger complex global predicament? What thoughts arise?
- What does it feel like in your body when you face the severity of the predicament? Is there a particular place you can feel your reaction(s)? Can you describe the sensation(s)?
- Do you notice a pattern that repeats itself when you are feeling overwhelmed? Do you get stuck in a particular train of thinking, a specific worry or fear? Does your pattern manifest in a repetitive behavior? Sometimes, these patterns can limit our full expression of life. If you discover a pattern, take note of it. As you journal more regularly and engage in the other exercises in this book, you can begin to discover patterns and interrupt them before they lead to rumination or hyperfixation.
- Complete the following sentences:
    - Right now, I feel . . .
    - When I feel [use the preceding answer], I notice [fill in the blank] sensation in my body.
    - I worry about . . .
    - What breaks my heart is . . .
    - Sometimes, I wonder . . .
    - I cannot . . .
    - I am . . .
    - One thing I will do to take care of myself today is . . .

Questioning Our Questions

"The times are urgent; let us slow down." This popular quote is brought to us by Bayo Akomolafe, from the Yoruba village in Nigeria. We know this invitation seems like the opposite of what we ought to be doing. How can we slow down as the world burns, people suffer, and species disappear?

In every step, we challenge ourselves to slow down, calm our nervous systems, and notice what feelings are arising in this frenzied state. Meaningful actions that are holistic enough to reorient our destructive systems will not come from our frenzy, panic, or urgency thinking. Solutions generated from this place of despair or desperation will be insufficient to address the complex predicament. Additionally, these solutions might cause unintended harm. Looking at our predicament from a relatively calm and clear state of mind, we can begin to see the bigger picture, the tangled web of overlapping crises. As we mentioned earlier, the climate emergency is not just about the climate. Even if we had a miracle technology that magically stopped global warming, we would still be overwhelmed by mass economic inequality, the annihilation of wildlife species, rampant ecosystem destruction, the overfishing of our oceans, threats of nuclear war, systemic racism, broken food systems, and plastic worming its way into every place on Earth, including the deepest tissues in our bodies. Let us repeat this: the climate emergency alone is not destroying our planet. The *conditions that caused* the climate crisis and kept us from addressing it for decades are extinguishing life and life-supporting systems.

If we rush into solution making with the intention of stopping the climate juggernaut, we miss the entirety of the complex predicament. Greenhouse gas emission is just one branch of a very big tree. We can myopically focus on this, cutting the branch or the entire limb. But until we address the roots of the tree, we are wasting valuable time, energy, attention, and the potential for transformation.[13]

In the words of Akomolafe, "Slowing down is not about getting answers, it is about questioning our questions."[14] What if, in our deepest yearning for lower greenhouse gas emissions and the end of fossil-fuel companies devouring our ecosystems, we have been asking ourselves the wrong questions? What if our narrow focus on climate has been a sleight-of-hand trick to distract us from redesigning our cultures to be truly life centered?

EXERCISE: GETTING BACK INTO YOUR BODY

If you are able, create some alone time, where you may not be interrupted for five or so minutes. Clear some space around you.

Stand up straight, breathe in deeply, and let out a long-exhale breath. Scan your body from the tips of your toes to the top of your head. What does it feel like to be in your body? Are there any sensations present?

Bring your attention to your right arm. For thirty seconds shake your right arm. Feel the sensations as you move it. Bring it up and down, left and right. Move it fast, slow, and fast again. Feel the flesh and muscles of your arm bounce and jiggle.

Now pause the movement of your right arm, and shift your attention to your left arm. Begin shaking your left arm. Bring it up and down, left and right. Move it fast, slow, and fast again. Feel the flesh and muscles of your arm bounce and jiggle. Notice how the movement of this arm feels similar to your right. How does it feel different? What sensations are you feeling?

Pause your arm movements and shift your attention to your right left. Begin moving your right leg. For thirty seconds bring it up and down, left and right. Move it fast, slow, and fast again. Feel the flesh and muscles of your leg bounce and jiggle. How does moving your leg feel similar to moving your arms? How does it feel different? What sensations are present in your leg? What does it feel like to have your flesh and muscles bouncing in different directions?

Pause the movement of your right leg and bring your attention to your left leg. Begin shaking your left leg. For thirty seconds, feel your left leg move and shake. Bring it up and down, left and right. Move it fast, slow, and fast again. Feel the flesh and muscles of your leg bounce and jiggle. How does it feel to move your left leg? What are the differences between moving this leg and your right leg? What are the differences between moving your legs and your arms?

Pause the movement of your left leg, and bring your attention to your torso. Begin shaking your torso, with a focus on your shoulders and hips. Expand your belly, contract your belly. Move your torso fast and slow and fast again. Twist and stretch your hips and shoulders. How does it feel to move your torso? In what ways is it similar to moving your limbs? In what ways is it different?

Pause the movement of your torso and bring your attention to your entire body. Begin shaking your whole body, in any way it wants to move. Up, down, left, right, in, out, fast, slow, anything else you can feel into. How does it feel to shake your body all over? In what ways is it similar to shaking the parts of your body separately? In what ways is it different?

Reflect for a moment, maybe in your journal, on what this exercise was like for you. How did it feel when you moved each of your limbs and torso? How did your full body want to move? Did any feelings come up? Did you notice any sensations? Taking note of the sensations in your body will help you better connect with your feelings in time.

Being with Our Discomfort

Curing our grief, rage, fear, or despair about the state of the world is not the point of our support groups and is certainly not the focus of this book. As we endure the Long Dark, these feelings will come back time and time again. This process is about helping us reconnect to our senses, bodies, and joy so that as the painful and heavy feelings return, we do not panic, avoid, or stuff. Instead, we can start developing a relationship with our feelings and welcome them as old friends who can stay for a cup of tea but are not permitted to take residence in us. This relationship starts with practices that help us accept our feelings, especially the ones that we have been told by everyone, from corporations to our parents, not to feel. All feelings are teachers, and some of the feelings that come from emotional states are more pleasant to experience than others. We prefer to use the term *heavy* for those feelings that weigh you down and make you feel like you will collapse under them and the term *painful* for those feelings that feel nearly unbearable to sit with.

There are no negative feelings; there are uncomfortable feelings. Viewing feelings as negative gives us reason to move away from them, instead of experiencing and processing them. We must learn to be with the discomfort that comes along with experiencing heavy and painful feelings, and instead of running away, we need to learn to lean into them and explore them. Heavy or painful feelings do not need to be cured but sensed and then processed when we are safe and resourced enough to do so. By doing this work, we create moments of spaciousness in ourselves, allowing us to deeply experience the enjoyable feelings as they arise, even as the outside world experiences ongoing disruption.

Our willingness to play into collective gaslighting may be indicative of our unwillingness to experience discomfort—whether that means facing uncomfortable truths, feeling undesirable feelings like anger or deep sadness, or making inconvenient changes to the way we live our lives. Psychologists call this inability to be with our heavy or painful emotions and our desperate urge to avoid or escape them as they arise *distress intolerance.* When a person exhibits low distress tolerance, the tendency is to shut down or deny these feelings—like those that might arise as we accept the realities of the global predicament. And in the consensus reality, shutting them down, denying, or avoiding them becomes the norm.

On an individual level, distress intolerance might have us neglecting our inner worlds. A few fellow Good Grievers shared how distress intolerance shows up in their lives. Taylor mentioned that when another parcel of public land is auctioned off to a fracking company, they immediately start planning how to interrupt the next auction. Dean explained that when he feels overwhelmed by the news, he numbs his feelings by reading article after article on the internet until he has lost track of time. Janelle shared that when a piece of critical legislation fails to pass, she stuffs her feelings by pouring a drink or shopping online.

Instead of acknowledging and processing our feelings, we push them down, down, down—and for a while, it works. We are able to stave off those feelings. But along the way, our refusal to feel can also impact us socially, perhaps making us irritable and grouchy or shut down and disconnected. The inability to process our feelings clouds our decision-making, blocks our imaginative capacity, and diminishes our capability to show up with an open heart. Eventually, our stuffed feelings will catch up with us in the form of illness, body aches, or a breakdown. The therapist, author, and somatic abolitionist Resmaa Menakem has said, "Only by walking into our pain or discomfort—experiencing it, moving through it, and metabolizing it—can we grow."[15] It is through the process of courageously attending to our emotional states that we grow and develop into mature human beings. Many of us try to create a better world at the expense of our own health and opportunities for maturation, and this fragmented approach mutes our efforts for the real, systemic change that many of us are working toward.

Individual distress intolerance leads to collective inaction. Many of us have trouble fully facing the severity of the predicament, and as a result, we have trouble identifying life-saving actions that we, as a larger community, could take to mitigate the worst impacts of the course we are on. We may glimpse into the heavy or painful feelings and fear that we might get stuck or be collapsed by them, but in the act of processing our feelings, we gather wisdom and insight. We are changed, becoming less reactive, impulsive, and fearful, which opens up whole new possibilities for imagination and collaboration. If we are not expending so much energy avoiding or stuffing our feelings, there is more energy available for cocreating new, less destructive pathways forward.

STRESS RESPONSES AND THE WINDOW OF TOLERANCE

As we are learning to increase our ability to be with discomfort, it is helpful to know about our stress responses, so we can move the discomfort and not get stuck in it. This process is about helping us to explore and reconnect to our somas—the embodied way we relate to the world. Our somas are the amalgam of our "body, mind, and spirit, plus our emotions, actions, experiences, relationships, cultural conditioning, and worldviews."[16] It is the whole picture of who we are, our internal world as it is shaped by and shaping the external world.

When faced with a stressor or potential threat, our somas engage mechanisms to help us survive. These stress responses, also called survival responses, follow a cycle: beginning (detection of threat), middle (action), and end (returning to safety).[17] Threats might be direct or existential—perhaps we have been affected by a major flooding event or not having enough money for food. Maybe we have had an argument with a loved one or been chased by a dog while out for a run. It is possible to have experienced an existential threat after reading the latest climate report, scrolling our Twitter feed, or watching a news broadcast focusing on the global impacts of the climate crisis. Furthermore, stuffing our feelings can elicit a stress response. For each of us, our stressors might differ based on a number of factors, including our instincts, lived experiences, and previous traumas. Our embodied stress responses are below the level of cognition, which means we do not think about how we will respond to a perceived threat; our somas make that decision and send orders to our brains to carry out.

After a threat or stressor is detected, a surge of biochemicals, like cortisol and adrenaline, are released in the body, resulting in "a cascade of physiological events, such as increased heart rate, respiration rate, and blood pressure; suppressed immune and digestive functioning; dilation of pupils and shifting of attention to a vigilant state, focused on the here and now."[18] This moves us into the action phase of the stress response cycle, where we are launched into a hyperaroused state (mobilization), wanting to run away or stay and fight the threat. Or perhaps our soma determines that we cannot fight or flee, and we are moved into a hypoaroused state (immobilization), where we freeze. The act of running or fighting, assuming we survive the threat, completes the stress response cycle by discharging the excess adrenaline and cortisol, allowing our nervous system to return to a regulated state. This state is where our somas are operating optimally—our mind is calm, and our body is able to maintain homeostasis. From this space, we can respond to stimuli and create connections. Dan Siegel, clinical professor of psychiatry at the UCLA School of Medicine and executive director of the Mindsight Institute, calls this state the *window of tolerance.*[19] We are in this state until we encounter another threat and respond with hyper- or hypoarousal.

When we are unable to complete the stress response cycle, we can get stuck outside our window of tolerance. If we are unable to run or fight when threatened, or if we are immobilized by the threat, the buildup of biochemicals remains in our bodies. We might experience overwhelm, anger, anxiety, or irritability if in a prolonged state of hyperarousal. Perhaps we get stuck in hypoarousal, experiencing numbness or dissociation or feeling withdrawn or otherwise shut down. If animals in the wild cannot fight or flee, they will shake or tremble to discharge the stress locked

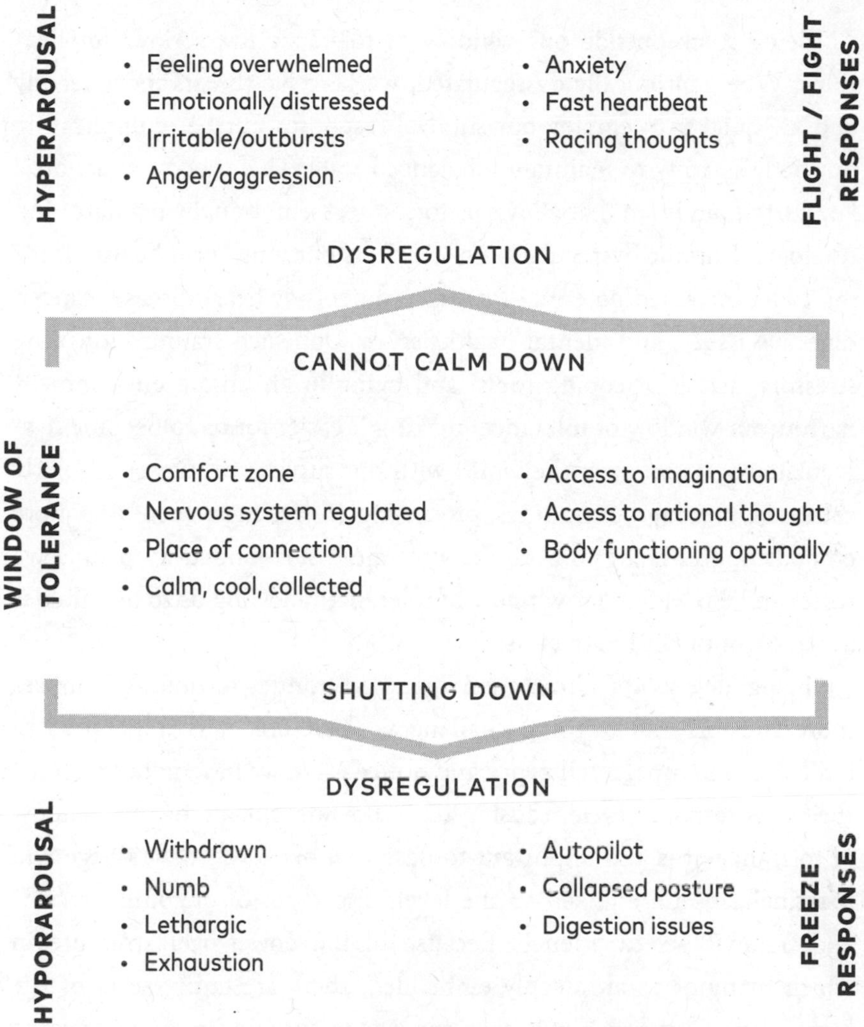

HYPERAROUSAL

FLIGHT / FIGHT RESPONSES

- Feeling overwhelmed
- Emotionally distressed
- Irritable/outbursts
- Anger/aggression
- Anxiety
- Fast heartbeat
- Racing thoughts

DYSREGULATION

CANNOT CALM DOWN

WINDOW OF TOLERANCE

- Comfort zone
- Nervous system regulated
- Place of connection
- Calm, cool, collected
- Access to imagination
- Access to rational thought
- Body functioning optimally

SHUTTING DOWN

DYSREGULATION

HYPORAROUSAL

FREEZE RESPONSES

- Withdrawn
- Numb
- Lethargic
- Exhaustion
- Autopilot
- Collapsed posture
- Digestion issues

A diagram of the window of tolerance and potential identifying features of the different arousal states.

up in their bodies after the threat has passed. We humans could do this too, or we could dance, yell, cry, express ourselves creatively, experience a big belly laugh, connect with others, or go for a run when we are no longer at risk.[20] Part of building our distress tolerance is learning to complete our stress response cycle, as it is safe to do so.

Being stuck outside our window of tolerance has serious implications. When chronically dysregulated, we perceive threats more readily and are quicker to employ our survival responses, further complicating our body's ability to maintain a balanced state. This process shuts off a key part of our brain that allows us to connect, emotionally regulate, and use logic. Chronic dysregulation also affects our long-term health, leading to an increased potential for such ailments as heart disease, cancer, digestive issues, and mental health issues. Unhealed traumas, ongoing stressors, a lack of coping tools, and being in an unsafe environment narrow our window of tolerance, making it easier for us to become dysregulated and approach the world with our survival responses regularly activated. Healing our traumas, processing our feelings, becoming more embodied, and being in a safe environment surrounded by people we trust can help widen our window of tolerance, allowing us to be nimbler as stress continues to affect us.

Being alive as the climate emergency continues to unfold promises more stressors and ongoing challenges. Therefore, it is imperative to build our emotional intelligence and notice when we have not completed the stress response cycle, causing us to operate outside of our window of tolerance. It is also important to note that our levels of privilege and marginalization are linked to the levels and types of ongoing stressors and traumas we experience. Because of the power-over structure in which many of us are deeply embedded, there are entire sects of the human population that have been exposed to chronic stress and ongoing systemic trauma—in some cases, for generations. We are not all arriving with the same window of tolerance, and this affects how we relate to each other as we move through the Long Dark. As we work to cocreate new paradigms that are life centered, we must work to correct this imbalance and extend compassion to each other and ourselves as we recognize that we are all in different places in our journey of healing, regulation, and connection.

STRENGTHENING OUR "BEING WITH" MUSCLES

Our ability to be with discomfort and move toward healing is intrinsically linked to our spiritual, psychological, embodied, and communal well-being. The impulse to turn away from heavy or painful feelings or from the work needed to complete the stress response cycle may seem like an effective coping mechanism in the moment, but resisting them as a strategy for life causes more harm than good. We already mentioned that being outside our window of tolerance can make us act in ways that are unpleasant to the people around us. We also know from Brené Brown's brilliant work on emotions that we cannot selectively numb our feelings.[21] Our attempts to mute the uncomfortable feelings also dampen our experiences of the more pleasant feelings. The richness of feeling becomes more available to us as we practice being with our inner worlds and building up this capacity as we might strengthen a muscle.

When we accept our emotions and understand them to be additional data points rather than judging them as negative or pushing them away, a funny thing happens: feeling our uncomfortable emotions can alleviate some of our suffering. It sounds paradoxical, but multiple studies have shown that when we use mindfulness exercises to accept painful emotions, it increases our ability to tolerate distress and decreases anxiety.[22]

As we practice being with our emotions, it may feel overwhelming, scary, or jarring. It might be helpful to learn that experiencing emotional sensations is a finite experience, lasting between ninety seconds and twenty minutes. Dr. Jill Bolte Taylor writes in *My Stroke of Insight: A Brain Scientist's Personal Journey* that after being emotionally triggered, it takes fewer than ninety seconds for the surge of chemicals to pass through the bloodstream before being flushed out. Staci K. Haines, in *The Politics of Trauma*, notes that the effect on our embodied state can last up to twenty minutes before this energy is discharged.

Feelings aren't always accurate; they are not logical or rational. Sometimes they arise because of a thought or belief that we have. Experiencing a feeling is always authentic, but perhaps the thought or belief from which the feeling arises is not a valid one. Feelings are data points that offer extra information about us and our situation, and gathering this information is a practice. Learning to be with our emotions is the process of pausing, observing, exploring, identifying, and, finally, choosing our reaction. Not all feelings require a reaction above and beyond experiencing them. Some do. Sitting with big, overwhelming feelings also takes some practice, and if you are like LaUra, it can be difficult to identify your feelings. There are many resources available to help us become bet-

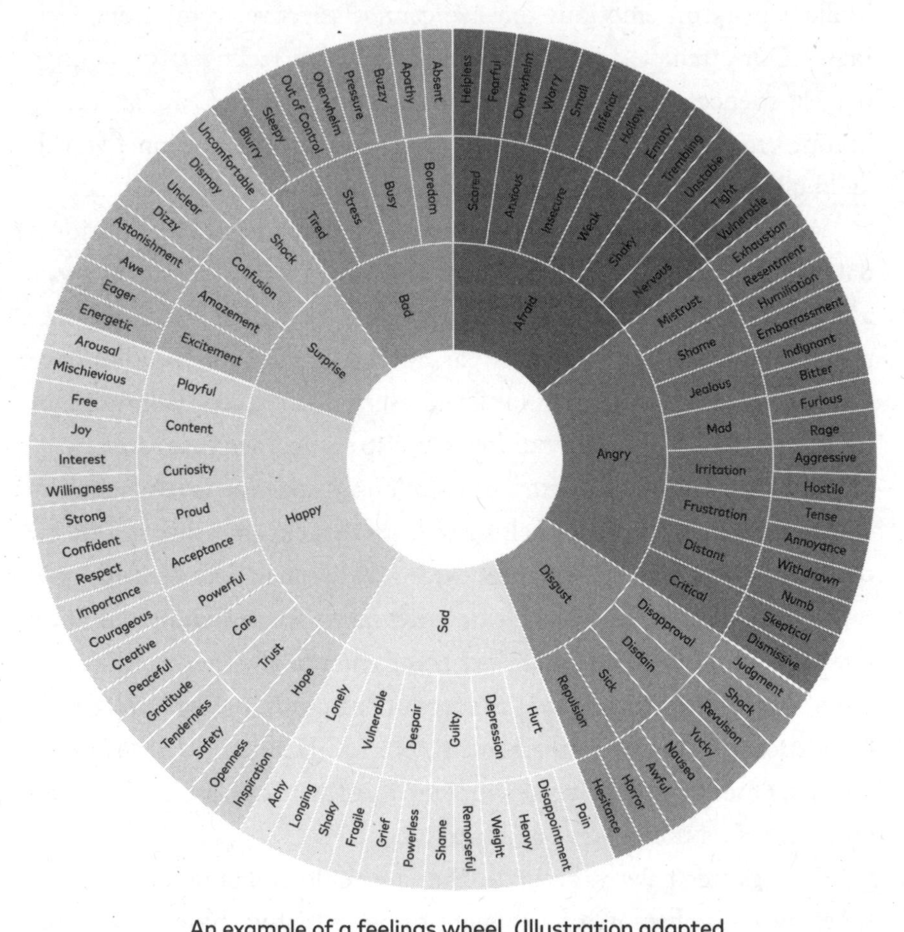

An example of a feelings wheel. (Illustration adapted with permission from Dr. Albert Wong.)

ter acquainted with our feelings. We recommend using a feelings wheel like the one above to help pinpoint feelings.[23]

Observing a more neutral or pleasurable feeling such as boredom, tiredness, joy, or excitement is a great place to enter into this type of work. Notice how your body feels when you are experiencing the neutral or pleasurable feeling. What does boredom feel like? What does joy feel like? This can serve as a baseline to return to especially as we are experiencing unpleasant feelings.

Accepting uncomfortable feelings means simply taking note of them and letting them move through you instead of creating a story around them or trying to ignore them. This work is not passive. It takes enormous courage and vulnerability to be with and process uncomfortable feelings. Over time and with the practice of feeling your feelings, your nervous system and your community will thank you.

Acceptance as Opportunity

> Acceptance does not mean surrender. It does not mean resignation. Acceptance means I am finally available to the entire spectrum of creative responses.
>
> —Trebbe Johnson, *Radical Joy for Hard Times*

It is time to get real about the emotional and physical impacts of the climate emergency. This direct and absolutely necessary invitation was first introduced to LaUra in a university lecture by the social science researcher and founder of the Adaptive Mind Project, Susanne Moser, in 2013, and it has stuck with her since.[24] Getting real and accepting the dire state of the world is grueling, gritty, painful work that might just trigger our stress and survival responses. While it does not become easier per se, we get braver at facing it just like we get braver about being with our uncomfortable feelings. We practice leaning into the hard realities and not rushing to escape or avoid. We can practice completing the stress response in healthy ways. It takes endurance, inner strength, and courage to hold space for distressing realities and emotions. That is why we begin our program with Accepting the Severity of the Predicament. If we cannot face the painful reality of our collective crises, how can we ever hope to change them?

At times, the grief that comes with accepting the predicament will feel unbearably heavy. It will carve us out, making us feel empty. And that is paramount because emptiness also means space—space to process difficult feelings and channel them into meaningful action and space to notice something different from the grief that has just been processed. When we allow ourselves to feel our grief instead of denying it, we give ourselves the option to transform it and be transformed by it. We can harness the energy of our grief and use it to collaborate, innovate, make meaning, and create new ways of being on this planet. Rather than weigh us down, our range of emotions empowers us to act in radical and inspiring ways.

Fellow Good Griever Liz Wade has shared that to accept the severity of the predicament is to "look at it, and not shy away, to feel all my feelings and hold them. Through this ability I can also hold my acceptance of the precious and sacred nature of life in its entirety. And with this strength and perspective I can also continue to work for change." She continues,

> Being part of GGN has taught me, in many ways, that this strength comes from our hearts. This strength comes from love. Love for the world, even as it is crumbling, burning, dying, and disappearing around us. Love for each other as we witness this together and hold our pain together. And love for ourselves. For we too are life, we too are precious, we too are sacred.
>
> We must remember that. I have worked to remember that.

Accepting the painful reality of the Great Unraveling does not mean we stop working toward change. As individuals and as a community, we must do everything within our power to ease suffering and hold a vision of what comes next. Acceptance is a directive to let go of outcomes and to open to a whole range of creative responses to the predicament. We act in ways that are congruent with our values, dignity, and purpose.

EXERCISE: BEING WITH UNCOMFORTABLE FEELINGS

Think back on a time when you felt overwhelmed by a painful or heavy emotion. (Note: If you are just beginning befriending your feelings, it is helpful to choose a relatively benign situation. You can work up to a more traumatic event, perhaps with the help of a healer.) The feeling you choose might be something small and uncomfortable that has been nagging, maybe for a day or two, maybe longer. Perhaps it was a careless thing you said to a loved one that you need to go back and feel your way through. Maybe a loved one said something to you, and you stuffed your reaction inside of yourself instead of feeling it. You may want to journal about this experience or simply create some spaciousness in and around you, as you do the following exploration. These directions are just suggestions. You know yourself, your body, and your reactions better than we do. You are the most important guide in this process. If a sensation becomes overwhelming, back off, take a break, and ease back into this exercise when you are feeling more resourced.

Take yourself back into that uncomfortable or painful moment. Remember what you felt at that time. Place your attention on a particular feeling that arises and soften to it. Explore it with curiosity. The emotion may intensify with direct attention, and that is OK. This sensation is temporary. As you place more attention on the sensation, begin to try to locate where in your body you can feel it. Is your chest or throat tight? Has your heartbeat changed?

Are your back muscles tense? Does your stomach ache? Is the sensation somewhere else? Multiple places? What does it feel like? Is it a burning, itching, tightening, tensing, or something else? Can you name the feeling associated with the sensation? As you feel your emotions and sensations, try not to judge them or create a story around them. Just experience and explore. With this focused attention, your body may want to have a release. Listen to what your body wants—a trembling, a cry, a scream or howl, an urge to punch a nearby pillow, or jumping up and down. Listen to your body and create space for any release that may be coming. Take a few long-exhale breaths through the release.

If a release does not come, that is OK too. The goal of this exercise is to remain curious about any feelings or body sensations that come and go. It is perfectly fine to be in the exploring stage for the duration of this exercise.

With or without a release, you may notice a change in the sensation. Is the feeling still present? Does it feel different in your body?

After about five minutes (you can lengthen the time of this practice as you do it more), go for a walk or shake your body out. Drink some water and practice long-exhale breathing. Do not rush back into your life. Create a few moments for quiet, calm, reflection, and integration.

Does your body feel different now, at the end of the exercise, than it did in the beginning?

Step 2

BE WITH UNCERTAINTY

If you're invested in security and certainty, you are on the wrong planet.

—Pema Chödrön

LaUra

One fall afternoon during my sophomore year, my mom picked me up early from school. "The bank is taking the house," she said. "We've got three days to pack our shit and get out. I'm getting a moving truck, but we can't take everything. So figure out what you're taking and do it soon."

Growing up around my parents' regular substance abuse and undiagnosed mental health issues, uncertainty was in the air I breathed. From an early age, I was all too aware of the unpredictability of my parents' behavior. One of my earliest memories was a domestic violence episode prompting my sisters and me to run across the street in our underwear to the protection of the neighbors. That night launched the unraveling of any stability I had previously had. What ensued were custody fights, visits from Child Protective Services, job losses, unpaid bills, and moving from home to home. I never knew whether my dad would be sober or drunk when he came to get my sisters and me in his pickup truck for visitation, when my mom would get sick of our beloved pet and remove them from our home, or when we'd get a knock at the door that someone was here to collect yet another car that had been repossessed. In middle school, my mom did not tell me she stopped paying on my saxophone, and they took that away too. For the rest of the semester, I was enveloped in shame every day during second period as I sat in band class with no instrument to play.

We moved to my house when I was in second grade, and for almost eight years, my home was my rock. My mother, a manic spender and compulsive liar, would often leave my sisters and me home alone day after day from the time I was a preteen. She purchased enough frozen burritos and corn dogs to get us through until her next return. Leading up to the foreclosure, she would make some comments here and there about the house but said she would figure something out. Between her tall tales and the absences, my sisters and I just assumed something would work out. Yet there I was, having to decide on the few objects I would take with me to our next stop—without knowing where that next stop would be. I spent the weekend in a state of shock, rotely sorting my life into a few plastic bins. I finished selecting which aspects of my life fit in the moving truck and which were to be left behind. Nothing felt real; my actions were mechanized. Our foreclosure date came, and just like that, the last piece of stability I had disappeared in the rearview mirror of the U-Haul.

After we lost the house, my mother eventually settled in with her boyfriend, who lived an hour away from my school and friends. I spent the remainder of my high school years staying with friends or on a fold-out mattress on the floor of my older sister's apartment. "Homeless" was not a person on the side of the road—it was me. It was my life. I was uprooted and unmoored. I responded to the instability in my life the way so many of us do: I shut down my feelings in an attempt to shield myself against the pain of uncertainty and loss.

I thought maybe once I made it to college and earned a degree, I could control my life and my decisions. That was the American Dream, right? A survivor of childhood abuse and neglect could become a free agent in a world full of possibilities. But as I dug into ecology, evolution, environmental science, psychology, sociology, and Buddhism, I started putting puzzle pieces together. The problems were social and ecological and incredibly complex.

Part of me mourns that the future I felt entitled to, especially after such a tumultuous childhood, is not available. I pine for stability, to have a place of my own to set down roots and to grow old with my wife. I yearn to see our nieces and nephews—whom we affectionately call our "niece-phews"—grow up and live long lives as they make meaning, experience joy, and build connections in their lives. There are days when I want to deny the state of the world, when staying in bed seems more appealing than thinking about where we are headed. I want an easy solution to the climate catastrophe, to our multilayered social issues, to ecocide. I want to meditate my way out of this despair. And eventually, I come back to the acceptance piece—I find myself back at Step 1.

Certainty Is a Privilege

When I take a step back, I realize that the illusion of certainty largely belongs to those of us in the Global North and, further, to those of us with the most privilege. For many of the world's most marginalized people, a stable future has never been promised. As I write, nearly forty-one million people worldwide are affected by famine,[1] and billionaires are jetting off to space during a pandemic. People living in coastal areas are losing their homes to rising waters and increasing storms, and it is estimated that the climate crisis will displace nearly 1.2 billion people by 2050.[2] Those of us in the Western world, operating under the dominant paradigm, have been under the power-over consensus trance while BIPOC communities, farmers, and fishers have been ringing the alarm bells, highlighting deep systemic issues. The scholar Farhana Sultana explains how the impacts of the climate catastrophe will impact those most vulnerable. She writes, "Historical differences position colonial and imperial countries at greater advantages over post-colonial and presently occupied countries. Colonial logics of extractivism continue through neocolonial and development interventions post-WWII. The ecologically unequal exchange between the Global South and Global North, ongoing extractive capitalism, the imperial structures of global trade, and domination in setting policies and ideologies—all work to maintain climate coloniality."[3] The instability created and perpetuated by the dominant paradigm has fallen disproportionately onto marginalized communities since colonization first began. What seems like collapse to those of us who have been sheltered has been an ongoing hardship endured by many across our planet.

The dominant paradigm operates by exploiting marginalized people so that a sect of the population can live more comfortably. Over time, this phenomenon expanded to the shadows, allowing many of us to be disconnected from the impacts of how we arrived here and how the impacts of coloniality and imperialism are still playing out today. As flooding, droughts, disease, and heat waves bring suffering and death to the impacted populations, we tell ourselves and each other, *That is so unfortunate, but there is nothing we can do about it.* Now the impacts of this way of being are undeniable and forcing us into a collective reckoning. Layer by layer the consensus trance is being peeled back, and the collective illusions are crumbling.

Uncertainty Is Reality

Regardless of our level of privilege, uncertainty and change are life's most fundamental truths. No one is guaranteed their next day on Earth. Even if we were not facing the disintegration of our biological, psychological, and social systems, uncertainty is reality. Accidents happen, illnesses develop, and our homes may be destroyed or taken away. A deep-time perspective can help us stay humble about the fundamental reality of change. On a planetary scale, Earth has been in transition for all 4.5 billion years of her existence. The Great Oxidation Event occurred 2.45 billion years ago. Its arrival caused extinction for the microbes that could not metabolize oxygen, yet it paved the way for multicellular organisms to exist.[4] An extinction event occurring 250 million years ago led to the decimation of 95 percent of marine species and 70 percent of land species,[5] which eventually allowed for the age of the dinosaurs. And of course, the extinction of the dinosaurs paved the way for mammals to dominate on Earth. And cosmically, stars burn out and are recycled into new galaxies. Our galaxy, the Milky Way, is about 13.6 billion years old, and eventually, our sun will burn out, and over time, all elements will be reconfigured into something else.

One of the hardest truths to hold is the understanding that everything is changing. It always has been. In *Parable of the Sower*, Octavia Butler writes, "The only lasting truth is change."[6] Each element in our bodies knows the cycle of growth and collapse. We are all temporary. Change is the air we breathe, the ether we swim in, the one absolute truth of this universe. Extinction is in our molecular makeup. Certainty has been an illusion all along. Instead of fighting it, how can we learn to embrace it with nimbleness, humility, and maybe even a bit of humor?

RIGHT RELATIONSHIP WITH UNCERTAINTY

So many of us in the dominant culture believe uncertainty is failure. We blame the individual for not making the right choices. Maybe they did not get the right college degree or the right job. They did not buy the right house or made poor decisions in family formation. It is their fault if they did not save enough money. And, of course, the pledge of care and safety continuously offered by the dominant culture is alluring. It speaks to us on a level beyond logic and gets right into our survival mechanisms. In the context of evolution, it makes sense that we would have a drive to minimize uncertainty. We want to know where our next meal is coming from and that we have a safe place to sleep. Unpredictability might threaten our safety or access to food. When we cannot anticipate what will happen next, feelings of fear and dread arise. It is unnerving and anxiety provoking.

One of the best things we can do to befriend uncertainty is to develop a relationship with it, to understand it not as a moral failure but as an intrinsic part of life. There is an old Buddhist parable about a farmer that comes to mind often. The story goes something like this:

One day, a farmer's horse runs away. When his neighbors hear the news, they visit to give their condolences.

"That's horrible luck," they say.

"Maybe," says the farmer.

The next day, the farmer's horse comes back with three more wild horses.

"That's amazing!" the neighbors say.

"Maybe," says the farmer.

The following day, the farmer sends his son to break the wild horses in the stable. One of the horses throws the son onto the ground, shattering the son's leg.

"Oh no, that's awful!" say the neighbors.

"Maybe," the farmer replies.

The farmer tends to the son's chores, doing twice the work in the fields. Later that week, military officers stop at the farmer's home to draft the son for war. Seeing him bedridden, they leave the son behind.

"How wonderful!" the neighbors say.

The farmer shrugs and answers, "Maybe."

Over the many years of famine and bountiful harvests, the wise old farmer mastered the art of nonattachment. He knew that life is unpredictable—and that more than any external tragedy, it is our own judgment that pains and limits us most. Buddhism, and other wisdom traditions, teach that if we want to live in right relationship with change and uncertainty, we must be like the farmer and let go of our attachment to the outcome. This story reminds us that uncertainty is neither good nor bad. It just is.

Arriving at a "maybe" response in the face of so much uncertainty requires living in right relationship with the unknown. This requires practices that help us find and maintain grounding throughout the tumult. It also requires that we heal old wounds and traumas that can prevent us from living in the present moment.

EXERCISE: EXPLORING YOUR EARLY RELATIONSHIP WITH UNCERTAINTY

We find this best explored as a journaling exercise and suggest writing out your responses to the questions below.

Think of when you were a child. When uncertainty creeped into your young life, how did the people around you handle the situation? When you encountered instances of uncertainty were the people around you scared? Were they excited? Did they react differently? How did these patterns help form you? Spend a few minutes exploring your earliest memories and experiences of uncertainty.

Now bring your awareness back to the current moment. As you read the previous section on uncertainty, what is your reaction? What feelings or sensations arise in your body? Note that it is OK if you do not feel any physical sensations. Do you have any impulses asking for your attention?

What is your current relationship with uncertainty. Reread the farmer parable and spend a few moments reflecting on it. Do you notice any reactions to the parable? Did it bring up any feelings or sensations?

POSITIVE DISINTEGRATION

> One is never afraid of the unknown; one is afraid of the known coming to an end.
>
> —Krishnamurti

We know there can be many conflicting feelings about what it means for our world to truly change. On the one hand, most of us yearn for a revolution. We want accessible health care for all, universal human rights and dignity, protection for rivers, oceans, mountains, and nonhuman beings, and mitigation of the worst impacts of the climate emergency. And we want this all *now*. On the other hand, we fear the end of the familiar and the void of the unknown. It is hard to imagine what "rapid, far-reaching and unprecedented changes in all aspects of society" looks like in the world. Because of our deep dependence of fossil fuels, things may get worse even if we are able to make the drastic changes needed to mitigate the worst impacts of the climate emergency. The thought can feel both terrifying and exhilarating.

We have a tendency to be resistant to change because it can feel destabilizing and evoke our survival responses. Yet change is fundamental for our growth on both an individual and collective level. Kazimierz Dąbrowski, a Polish psychologist, developed a theory called *positive disintegration* to describe this phenomenon.[7] On an individual level, positive disintegration is the process of questioning and eliminating old beliefs and values so that we can adopt new ones that benefit our personal growth. Buddhism and systems scholar Joanna Macy applies this idea to the systems we live in. For us to cultivate truly life-sustaining systems, we must question and dismantle our existing systems. This is part of the Great Unraveling. We are being ushered into a collective positive disintegration, where we let go of the ways of being that are no longer serving life. We can see the Great Unraveling as an opportunity to reorient a very sick paradigm. One afternoon, while we were sipping tea in Joanna's living room, she said, "The faster this society collapses, the better our odds of survival." We know what a painful statement that is, and we also agree. The dominant paradigm must unravel and, in doing so, make room for new ways of being.

Nothing grows infinitely. The natural rhythm of the universe is expansion and contraction. We see it in the formation of galaxies and stars, the rhythm of ecosystems, and the population growth of species. Our young human species got lost in the mandate to grow, grow, grow, without understanding that there is a time to contract, to slow down, to collapse, to decay. Disintegration is necessary and natural. Death is part of the cycle of life. When we allow the old life-destroying structures and ways of being to fall away, we make more room for new ones to take root and grow. And we need new, life-centered ways of being to root and grow.

We can approach positive disintegration with more openness and curiosity when we embrace a continuous willingness to get lost. If we know our current ways of being are killing everything around us, how can "we lose our way, generously"[8] as Bayo Akomolafe suggests? If we are unwilling to get lost, if we are unwilling to question our assumptions about the way things are, how can we ever expect to create new ways of being? We get lost to find new ways. We get lost to identify new worldviews or to help us remember ancient worldviews that have been forgotten.

Slowing down and losing our way allows us to notice new ways that are springing up that we have not yet acknowledged. These new ways are openings that are dancing on the edges of our understanding and perspectives waiting to be witnessed. The late meditation teacher Jerry Granelli first taught me this concept of *openings*—opportunities for novel thoughts, actions, or ways of being to emerge that we may have been too distracted or biased to notice before.[9] Openings are a third way, the open door that arises, a portal that appears, if we can tolerate the tension and unknown and become present with what is. We are often doing too much, going too fast, or not embodied enough to notice the openings as they appear.

But we cannot rush the openings, just as we cannot rush getting lost. The disintegration process allows space for the undoing work. This is a time to embrace the messiness, to make room for the vast uncertainty that is the new normal, and to endure the tension that comes with being with the uncertainty. We must mix it up and rely on our creativity, courage, and connections throughout this process. Only when we are truly lost can we start to identify the next best step forward.

The Wandering Nerve

We humans are a risk-averse species. Over millions of years, our ancestors survived by avoiding and minimizing change and uncertainty. By seeking relatively low-effort, reliable sources of food, warmth, and shelter from the elements and safety from predators, our ancestors endured long enough to eventually create us, modern humans. To satiate our needs for security and certainty, we built a world designed for stability, complete with air-conditioned buildings, mass-produced foods in convenient supermarkets, fences to keep "wild" animals out, and emergency services we can summon simply by picking up the phone.

Over time, we have also evolved complex survival mechanisms, many of which are centralized in the vagus nerve. The vagus nerve starts at the brain stem and wanders its way down through much of our torso; hence its nickname, "the wandering nerve." This nerve sends key information between many of our internal organs and our brain. Nearly 80 percent of the information is sent from the body to the brain, with only about 20 percent coming down from the brain to the body.[10]

The vagus nerve is the primary nerve in our autonomic nervous system, which controls our involuntary bodily functions like our heart rate, blood pressure, breath, and even our facial expressions. Polyvagal theory, developed by Stephen Porges, teaches that our nervous system has a range of states of being, all of which are affected by the vagus nerve. (For the purposes of this book, we are simplifying the anatomy of the vagus nerve and its intricate relationship with our emotions and survival/stress responses. We highly recommend the work of Dr. Stephen Porges and Deb Dana for a more nuanced exploration of the wandering nerve.)[11] When we are feeling calm and safe, our nervous system is regulated; we are within our window of tolerance. The vagus nerve is in a rest and digest state, focusing on metabolizing our food, conserving energy, and repairing the gut. We are relaxed and able to connect with those around us.

The stress responses we discussed in Step 1—hyperarousal (fight or flight) and hypoarousal (freeze)—are controlled by the wandering nerve. Again, our nervous system does whatever it deems necessary to survive, often unconsciously. If we survive a stressor or threat, but fail to complete the stress cycle, the response chosen by our soma becomes conditioned in our bodies. Moving forward, each and every time our safety is threatened (whether the danger is real or imagined), we respond in the conditioned way. In other words, our conditioned responses generalize to many situations and not just the initial instance. Similarly, as we survive traumatic events, our trauma responses also become conditioned, survival responses until they are healed. If we do not participate in healing work, these responses limit the ways we engage with life and those around us. While they helped us survive a stressor, threat, or traumatic event at one point, they can keep us closed off from new opportunities or new ways of reacting to stress or trauma.

Our survival responses evolved at a different time for a very different set of circumstances. They must have been helpful to our ancestors when faced with a giant twelve-foot-tall short-faced bear, but today, our threats look different. Not all our modern threats are life-threatening in the moment; many instead endanger our sense of reality, certainty, and our projections of future safety for ourselves and those closest to us. When we watch a news report about the permafrost thawing and unleashing unknown and potentially dangerous microbes into our humanscapes or wildfires ravaging homes, our stress responses become engaged. When gas prices increase, or we do not know if we will make our monthly mortgage payment, our security feels threatened. And when new IPCC reports are issued saying our window of time to mitigate the worst impacts of the climate emergency is closing, we can collapse under the projected threat of our collective survival and operate in our daily lives stuck in our trauma responses.

When we sit with the existential dread of having to change everything about the dominant paradigm's social makeup to survive the Long Dark, it is almost too much to cope with—and this compounds the trauma many of us already have from our life circumstances, from systemic oppression, or from impacts of the climate emergency. For those of us who have experienced wildfires, food shortages, storms, or other traumatic events, the threat of the climate crisis is not theoretical or existential. It is experienced, and the body will relive it over and over again and act out its stress responses until the trauma is healed and a new response can be enacted.

As we encounter more threats to our safety—whether existential or experienced—our compounded traumas trigger our conditioned survival responses, and we begin operating in hyper- or hypoarousal or a combination of both, which can be more damaging than helpful. Living in a state of hyperarousal demands significant energy from us and causes our bodies to constrict and disconnect. We become more sensitive to perceptions of threats, more prone to anger, violence, and self-protective actions. If we are stuck in a state hypoarousal, we are frozen, numb, and detached from the present reality. The reactions of the vagus nerve occur before we have time to rationally think and decide how to respond. We cannot reason our way out of these responses. Our bodies make the response decision without our conscious awareness, and our conscious awareness shapes a story around our body's response.

Healthy regulation requires us to heal our conditioned survival responses and teach our bodies how to return to a state of homeostasis. To do this, our somas must experience a level of relative safety. We use the word *relative* because, as the trauma expert Peter Levine explained in an online talk in 2022, people caught in war-torn countries (and, we will add, those directly affected by the climate catastrophe) might not have a baseline of safety to return to.[12] Instead, we must establish relative safety as a precursor for regulation. This is further complicated by living in a traumatizing culture during the ever-unfolding climate emergency, making it more essential we begin healing our conditioned responses now.

THE IMPACTS OF UNHEALED TRAUMATIZED SOMA

Our relationship to uncertainty is deeply impacted by our experiences of traumatic events. Events that threaten our safety, belonging, and dignity activate our survival mechanisms.[13] Trauma occurs not because of a specific event but because of the lasting impacts of an overwhelming event and not having the appropriate resources available to help us move through the experiences. If we do not heal our trauma, it lives in our somas and informs the quality of how we show up in the present moment.

The grief worker, therapist, and soul activist Francis Weller describes trauma as a "rough initiation."[14] Typically, an initiation is the beginning of a rite of passage, a time of soul expansion in which the old form is left behind, and the dying away provides space for a new way of being. Traditional initiations are usually deeply communal practices, held and contained by our community members. When done with intention, awareness, and care, the individual's transition or maturation becomes a gift for the whole community. For this reason, Weller calls initiations "contained encounters with death."

Trauma, however, is a rough initiation because it does not have the support or guidance of community. During or after a traumatic event, we are often left to cope in isolated, individualized ways. Weller explains that, in this way, instances of trauma are "uncontained encounters with death." To survive our traumatic situation, we fragment, turning off parts of ourselves and constricting so far into our armoring that only a shell is visible. Part of the healing journey is to open the constriction, uncover the fragmented part, nurture the wound, and contain it by reintegrating back into the collective.

There are many types and levels of trauma that can overlap and influence one another, making it very complex individually and collectively. We may think we do not have unresolved trauma in our somas, but as members of a power-over structure, we are all arriving with our own experiences of fractured safety, belonging, and dignity, which lead to trauma. If we do not heal these wounds, every aspect of our lives, from the way we perceive the world to how we care for ourselves and treat others, is implicated.

A relatively new type of trauma—climate trauma—is now coming into focus. Climate trauma is a complex phenomenon brought about by climate catastrophe.[15] Climate breakdown involves multiple levels of our social and biological systems. Its impacts reach our entire biosphere, all types of species, and our sociocultural systems. It affects our jobs, families, and communities and exacerbates and is exacerbated by other traumas, such as intergenerational and historical traumas. And as we have mentioned, climate trauma also impacts our individual psyche. As the climate emergency rains increasingly severe impacts upon us, it will increase our instances of trauma, threatening to overwhelm our nervous system if we do not make room to slow down, heal, and reconnect. Furthermore, by receiving constant reminders from news outlets, social media feeds, and scientific reporting, our survival responses will be regularly engaged, and we will experience ongoing disruptions, resulting in present, future, and existential anxieties and dread.

TYPES OF TRAUMA[16]

- **Type 1 trauma:** Single-incident traumas or big *T* traumas, such as experiencing a sexual assault or a car accident, being

robbed or mugged, or living through an extreme weather event
- **Type 2 trauma:** Repetitive or complex traumas; for example, long-term domestic abuse, living in a war zone, long-term misdiagnosis of a medical issue, abuse or neglect from parents in youth, living with chronic medical conditions, or repeated sexual abuse
- **Secondhand trauma:** Bearing witness as someone speaks about their direct trauma and not processing its impacts on the listener
- **Historical trauma:** Inflicted by a dominant population onto another population, such as racism or genocide
- **Collective trauma:** An event or series of events affecting an entire population, like flooding, fires, or the impacts of segregation
- **Intergenerational trauma:** Trauma passed down from generation to generation genetically or from living with trauma survivors
- **Little *T* trauma:** The everyday losses or experiences that can add up if not healed, like a relocation or job loss
- **Climate trauma:** The complex trauma brought about by the climate emergency

The aversion to grief, pain, and the expression of heavy and painful feelings pervading the dominant paradigm compounds our trauma and keeps us from authentic connection. We hide our pain, which can manifest as shame. We may reach out to loved ones only to find they may not have the emotional skills or bandwidth to support us. It is possible that our loved ones carry unhealed traumas that become triggered as we try to heal our own. If we are unable to be with our own painful feelings and experiences, we cannot be with others' either. Many of us are left to process trauma alone, with few emotional tools to do so. When trauma goes unhealed, our relationship with uncertainty becomes one of fear and anxiety, triggering our flight, flight, or freeze responses.

Trauma left unhealed has us existing in fragments. To survive, we create an internal disconnection, desensitizing us from our own emotions, sensations, and intuition. Trauma teaches us that our bodies are unsafe to be in, so we cope by dismissing, disassociating, numbing, or constricting or by a whole slew of other techniques. By existing in our survival responses, disconnected from ourselves and other people, we become easier to manipulate and control. If we are living in a somatic space where our life recurrently feels threatened, we will act out of desperation to dispel the threat. This might include grasping onto the false promises offered by corrupt leaders, which we are witnessing play out as political polarization increases in the United States. The far right is using the chaos of these times to drive wedges between communities and increase fear. Those we love, or perhaps we, ourselves, may be pawns in these attempts until we heal, bringing back our sensations, connections, and intuition. We will discuss more about biases and death anxiety in future steps, but it is important to note that when we feel threatened, we will do just about anything to survive and numb the pain, even if it acts against our collective best interests.

The buildup of trauma in our somas also erodes our capacity for empathy.[17] This shutting down of empathy promotes Othering and causes harm, as we detach from the consequences of our actions. Our survival responses, the internal mechanisms that we used to survive, were valuable to help us at one time but severely limit our ability to experience joy, be in the present moment, feel centered, and connect on all levels. If we are not healing our wounding and regulating our nervous system, we will be continually tossed about in the ocean currents of the uncertainty, chaos, and pain of the Long Dark. To be fully alive, connected, grounded, and inspired, we must learn how to regulate our nervous systems and heal our wounds.

UNHEALED TRAUMA

Unhealed trauma is limiting and can do the following:

- Keep us disconnected, isolated, and fearful of each other
- Keep us distracted, addicted, and waiting for perfection before we take meaningful action
- Block us from accessing our imagination and believing things could be different than they are now
- Have us thinking we deserve our current negative situations and that there are no other ways
- Make us easier to control and manipulate
- Keep us stuck living in systems that are not life centered, allowing us to continuously perpetuate these toxic systems

THE SOCIAL ENGAGEMENT SYSTEM

Our ancient survival instincts tell us to put up a fight, run away, or shut down, when it might be more beneficial to stay calm, think clearly and, most importantly, connect with others. These must become our new survival skills.

To access this capacity, we must learn to regulate our nervous system by activating our *social engagement system.* Like fight, flight, and freeze, the social engagement system also operates via the vagus nerve, but unlike our survival responses, our social engagement system *enhances* our ability to connect with others rather than pushing them away or running from them. When we feel secure, the vagus nerve helps us create appropriate facial cues, like smiling. It improves our verbal abilities and our hearing and helps messages from the brain to the body travel more quickly[18]—all of which enhance our ability to communicate and foster connectivity.

Our social engagement system works in a loop: a calm nervous system helps us create safe connections with others, and safe connections with others can also calm our nervous system. This is why sitting with a wise friend, a trusted healer, or a support group can feel so medicinal. As the author, self-care advocate, and coach Cheryl Richardson says, "People start to heal the moment they feel heard."[19]

In this time of rearranging, we could be triggered again and again by nearly anything if our nervous system is continuously dysregulated. Thankfully, we can see our individual triggers as teachers. They point to an unhealed wound within us. If we can notice our triggers and explore them with curiosity, we are invited to place our attention on the situation or moment that elicited that particular response. We can investigate how we are feeling. What triggered us? What was our reaction to the event? What is the underlying wound the trigger is pointing to?

It is in our collective best interest to begin healing our traumas and learning techniques to regulate our nervous systems to get the social engagement system back online when we are launched into dysregulation (we suggest some exercises and practices in the next section). Individual daily practices can help us process, regulate, and find grounding amid the uncertainty. Cultivating or joining a community of practice made up of attuned individuals can help contain potentially traumatizing events. If it is not possible to regain operation of our social engagement system, we can at least start to recognize how our trauma responses and dysregulation cause us to react in ways that are not in alignment with who we want to be.

Practices for Grounding

> Practice is what gives us the clarity, presence, intelligence, and empathy necessary to be fully given over to anything we do, including activism.
>
> —Terry Patten, *A New Republic of the Heart: An Ethos for Revolutionaries*

We are in a liminal space, stuck between worlds, one collapsing and one asking us to usher in its birth. What comes next is unknown. Timeless Buddhist wisdom conveys that we are always in transition. Impermanence is a core guiding principle of reality. In *When Things Fall Apart: Heart Advice for Difficult Times*, Pema Chödrön reminds us, "To be fully alive, fully human, and completely awake is to be continually thrown out of the nest. To live fully is to be always in no-man's-land, to experience each moment as completely new and fresh. To live is to be willing to die over and over again."[20]

But groundlessness can be incredibly disorienting and terrifying. And while we are actively inviting disruption to BAU and getting lost gracefully, it would be helpful to maintain some clarity of mind and spirit. Aimee once said that she feels like the rug is constantly being pulled from under her feet. "I guess I better learn how to dance while falling," she offered. To dance while falling is a wonderful metaphor for practicing being with groundlessness. We anchor ourselves not to the ground beneath us but to the present moment, our values, our communities, and to the ever-unfolding process. And we can dance while doing it.

We are always practicing something, whether it is intentional or not. Our attention, life force, and presence are always being invested some place. If we are not present or aware of where we are using our energy and attention, we might feel regularly victimized by life's many challenges. Cultivating and nurturing intentional practices allow us to add grounding, presence, and connection in the moment. Establishing grounding practices throughout the Long Dark can help us maintain a clear mind and presenced soma. They will help us maintain agency over our reactions, and therefore minimize harm, as our survival responses will be regularly engaged. Because we know more difficulty is on the horizon, it behooves us to begin a regular grounding practice right now.

The late author and integral teacher Terry Patten taught that a practice is "not just *what* [you are] doing, but *how* [you are] relating to it (and to everything), and *why*."[21] When we are practicing, we are in relationship with the larger world, the moment, and ourselves as we bring mindful awareness to our sensations, attention, and desires. Sarah Jornsay-Silverberg, a fellow Good Griever, shares the what, how, and why of her practices:

> I cultivate the time, space, and discipline to engage in my daily practices not out of obedience to past resolutions or because I believe that doing them will earn me some future reward but because by committing to daily rituals that bring me to life, I am deciding that I am worthy of being brought to life every day. Following a routine gives me permission to take at least a moment each day to touch joy, presence, and my own electric aliveness. Choosing to be present to my precious human experience matters, especially in a world that is coaxing me toward numbness and distraction at every turn. These practices are a tether to the world within and around me; they remind me that it is a gift to awaken every morning to a body that can bear witness to these times of great transition.

Inspired by a map of individual practices by Patten,[22] we believe practices can include many realms of our being: spirit, mind, body, and inner work (or emotional work). In the table below, we offer suggestions for a regular practice in each of the categories.

TYPES OF GROUNDING PRACTICES	SUGGESTIONS
Spirit	Ritual, ceremony, spending time outdoors, prayer, contemplation, meditation or mindfulness
Mind	Studying, experimenting, critical thinking, building self-awareness, reading, engaging in meaningful conversations, journaling

TYPES OF GROUNDING PRACTICES	SUGGESTIONS
Body	Eating a healthy diet, practicing yoga, walking/jogging, performing intuitive body movement or intentional breathing exercises, drinking water, getting quality sleep, having sex
Internal/Emotional	Noticing and processing feelings, identifying emotional sensations in the body, identifying and exploring triggers, healing personal and collective traumas, creating communities of practice

We never become masters of living with uncertainty because uncertainty goes against our biological programming for safety and survival. We can, however, be in right relationship with the unknown and grounding practices help us arrive there.

Creating the Conditions for Emergence

Systems are all around us and make up every aspect of the world we inhabit. A system is a complex entity made up of smaller individual parts. The country of Canada, a flock of geese, the cells of a tree, a conversation among friends, and the US stock market are all examples of systems. While each system is made of smaller parts, the whole system is greater than the sum of its parts. Systems theory studies the relationship between the individual parts of complex systems. Our global predicament is one outcome of interacting systems. We mentioned this effect in the introduction and Step 1, when we discussed the overlapping nature of climate, human rights, capitalism, ecocide, and more to create the larger predicament.

We must alter everything from our global systems down to our individual actions if we are to cocreate truly life-sustaining ways of being. We can feel powerless to effect the change needed, but this is when it is crucial to remember that relationships matter. Our individual actions, when plugged into the collective, are needed for this transition. While we cannot control the outcome of our actions, we can create the conditions for new ways of being to emerge along the way.

Emergence is a key part of systems theory, as well as philosophy, art, science, leadership, and activism. Emergence is a phenomenon occurring when the parts of a system interact and create an outcome not predicted by the sum of the individual parts. In other words, it is the magic that happens when the many small parts of a system generate something new and unexpected through their relationships. Emergence is the X factor, the spiraling upward from an individual action to a collective phenomenon, a game of telephone with unpredictable results. A classic example of emergence in nature is a flock of starlings. Together, hundreds of starlings can fly in perfect unison, collectively changing direction on a dime to create shimmering, twisting formations in the sky. This seemingly supernatural formation is called a *murmuration*, which scientists discovered is patterning created by each individual starling quickly tracking and responding to its seven closest neighbors.[23] They are in coherence with each other, acting as a whole organism while flying through the air. When one moves, the next starling mimics the movement, causing rapid waves of twists, turns, and dives to ripple throughout the entire flock. This is the power of individuals working together in relationship to help something new come into existence. We are limited as individuals, but when we are part of a collective, we can share responsibility for birthing a life-centered future.

While we are adapting to a climate-changed world, emergence creates space for the unexpected to come through while honoring that it is not on anyone of us to save the world. Another emergent strategy principle from adrienne maree brown is "small is all"; in other words, the larger world is a reflection of small, everyday actions and events.[24] brown teaches about fractals, an important aspect of systems theory, to illustrate this principle. The Fractal Foundation defines a fractal as "an infinitely complex pattern that is self-similar across different scales."[25] A great example of a fractal pattern is the shape of a spiral. We see this pattern repeated in our fingertips, in snail shells, in galaxies. If you use a microscope and look at the micro level, the spiral just keeps getting smaller. If you use a telescope and look up to the macro level, the spiral gets vaster and more expansive. The spiral is without a starting or ending place. The pattern repeats itself at different scales.

Similarly, our small actions repeat themselves at scale in the larger collective. Our culture and social systems are made up of decisions that we humans buy into and perpetuate. These decisions are collectively agreed upon and carried out by individuals. When the world's plethora of problems and their unknown outcomes threaten to overwhelm us, when it feels like we are not moving fast enough and cannot find a pathway out, we are invited to return to the individual level. If we focus on doing our own inner work, deconstructing cultural norms and showing up in collective spaces, there is no saying how fast we can create new ways of being.

There will be times when our work will never feel like enough to change the dominant culture's destructive ways. It might feel like *I am only one person; I can only do so much*. But remember, small is all. What we do on the micro level can ripple into the collective.

Once we let go of the idea that we, individually, are responsible for saving the world, we are free to live right here, right now. We tend to our internal worlds, knowing that it directly changes how we view and respond to the external world. Our bodies can become an anchor to the present. The world is born anew as we slow down to fully absorb the enchantment of a cat's meow, the setting sun, or a cup of tea.

The Seed Bank

> The opposite of faith is not doubt, but certainty.
>
> —Anne Lamott

The dominant paradigm fosters a belief that things are worth doing only if we can guarantee results. But the truth is that results are never promised. Another of the emergent strategy principles outlined by adrienne maree brown is that "change is constant," and so we must "be like water."[26] Water does not need to know where it is rushing to next, what river or ocean it will end up in. It flows without anticipation.

We are continuously practicing letting go of the outcome of our work. To wait for a guaranteed outcome is limiting and keeps BAU trucking along. We are each being asked to practice nonattachment and do the work that is meaningful and joy filled for us. The process of doing the work to connect, heal, and repair is enough. We talk more about meaningful actions in Step 10. But for now, what gives your life meaning and joy? What work would you be doing if you did not have to guarantee the results ahead of time?

We need faith that movements beyond what we are aware of are happening. We maintain faith that if we do work that is collaborative, healing, regenerative, and relational and do it without expected outcomes, we are laying the groundwork for a better future—even though we may never see that future become a reality. We may not be able to control the outcomes of the climate crisis, but we can plant the seeds for a future built on love and focused on relationships. At worst, our seeds will never bear fruit, but we can live knowing we have done the right thing by cultivating an incredibly meaningful and connected life. At best, those seeds will grow a future that is beyond our wildest ideas of what's possible. It is not for us to know the outcome; it is our sacred duty in these times to create the conditions for the magic of emergence to take over.

Think of a seed bank in the natural world, with a variety of seed types dispersed all throughout the layers of soil. Imagine, maybe, that it is located on a riverbank. We do not know what will emerge on the riverbank from year to year, but we know there are many types of seeds lying dormant just waiting to sprout. The seeds that germinate depend on the conditions in that particular area. Was there more rain this year? Maybe more varieties of plants that love rain will grow. If it is drier, perhaps we will see more drought-tolerant plants sprout. How deep are the seeds buried? Are there seeds waiting to sprout until the bank has eroded a few layers? As conditions of the seed bank change, different seeds emerge.

Our collectives are seed banks. Our actions till the soil and plant our seeds: love, justice, diversity, compassion, and cooperation. We nurture the conditions for our seeds to emerge, and when the time is right, they will germinate and eventually shoot above the soil. What seeds are you planting in your life?

The Ever-Changing Present

"Change is coming, whether you like it or not." I, LaUra, have known this truth since I was an undergraduate student studying biology, environmental studies, and religion. It is one thing to study it, to see patterns and trends, and to know we are not moving fast enough to avoid the worst impacts of the climate crisis. It is another to see it unfolding in real time: the pandemic, the racial reckonings, the internal collapse of my friends and family, the fires and storms, the heat waves, the increase in violence and polarization, and of course the collapsing ecosystems and mass species extinction. It is wholly another to know that we must experience this rearrangement. These realizations can be enough to shut us down, to constrict us, and to send us falling into our trauma responses.

When I sense my anxiety spiking and am seeing only through the lens of catastrophe, I remember James Baldwin's wise words: "No one can possibly know what is about to happen: it is happening, each time, for the first time, for the only time."[27] I cannot predict the future. There is no set destination. There is always space for creation and visioning. There is room for imagination and excitement. Instead of feeling like all our global efforts are lacking, and we are destined for the worst types of chaos, I focus on the small actions that I, as one individual, can take. I focus on what brings meaning and enhances connection and joy. I continue my work without striving for perfection or a specific outcome.

If we only start something when we know we will succeed, we will never be brave enough to try new things. Our faith in our ability to change, to try new things, is what allows for the greatest inventions to happen. So often I hear people say, "There is no hope, so why try?" And it becomes a self-fulfilling prophecy.

To really practice being with uncertainty is to come back, time and time again, to the notion that we do not know how things will unfold. We are a young species who has just barely evolved here on our home planet. We have limited perception, and while we have learned a lot about life and life-supporting systems in the past few hundred years, there is much to learn and even more to experience. To be in process with. To witness.

Being with uncertainty is to practice being with the questions. To stop trying to force things. To relax the slack on the leash. To free up the mental energy that certainty demands. Being with uncertainty is a request to slow down, assess, and reassess the type of relationship we want to have with the more-than-human world. Being with uncertainty is about flexibility instead of rigidness. It is about going with the flow instead of clinging to the bank of the river. It is about staying curious, being open, and experiencing awe, as often as possible. When we stop clinging to what we know, we open to imagining what previously seemed unattainable.

Fellow Good Griever Lou Leet shared that letting go of control and being with uncertainty was a type of freedom for her. She said,

> I could stop constantly worrying about the future. I was left with more energy to do the things that I chose to take on. . . . It diffused the desperation I felt that others should see things the way I saw them and feel the way I felt about them. It left more space for me to do my inner work and bring it to the outer world when

I was ready. It made room for play. It made room for rest. It made room to see the beauty that exists—not a little bit of beauty but so *much* beauty.

The past has already happened, and the future, despite all our predictions concerning climate breakdown and positive disintegration, is not set. The endpoint never comes because the process is here now, constantly unfurling.

Step 3

HONOR MY MORTALITY AND THE MORTALITY OF ALL

> Should you fear that with this pain your heart might break, remember that the heart that breaks open can hold the whole universe. Your heart is that large. Trust it. Keep breathing.
>
> —Joanna Macy and Chris Johnstone

LaUra

I live in a world full of ghosts. They linger among single-use plastics, gasoline-powered vehicles, coal trains that catch me at the railroad tracks on my way to the box store, and television commercials. These ghosts find themselves in fast fashion, concentrated animal-feeding operations, and our monocultures. All things sustained by fossil fuels must be rearranged, reenvisioned, and most of them left behind in this paradigm. We are going to lose a lot of what many of us in the dominant culture take for granted. This truth is present with me every day, and I struggle with the tension of participating in this destructive culture. The awareness haunts me, and until I created GGN, it was suffocating me.

The dominant paradigm's time has come. The illusion provided by fossil fuels that we can maintain infinite growth on a finite planet is shattering. Yet we are addicted to this way of living. It is hard to imagine anything but what we have right now, and many of us do not want to. Human psychology makes us believe that tomorrow will be like today, even when we are presented with information to the contrary. We believe someone will save us, the problem will not affect us directly, or that it simply cannot be happening. But when all these biases crack open and the severity of the predicament forces its way through, the reminders can flood our bodies with complex emotions as the dominant culture continues BAU, eliminating any opportunity to restore a habitable planet. The dominant paradigm is unsustainable in the truest sense.

We live on a planet with finite resources, some of them renewable and many not. Nonrenewable resources like oil, coal, and natural gas cannot be replaced after they are used up. Water, air, and forests are thought to be examples of renewable resources that replenish over a short amount of time. Because the global economy is largely based on extraction and consumerism, we have mechanized the gobbling up and spitting out of a significant portion of Earth's resources. For example, by late July of each year, humanity uses up our global allotment of resources for the entire year, throwing us into "overshoot."[1] Simply put, humanity's demand for resources is using them faster than the ecological capacity to regenerate said resources. To add a significant layer of complexity, Earth Overshoot Day varies from country to country, depending on the level of consumption.[2] There are even some countries that do not even have an overshoot day.[3] Those in the dominant paradigm devour our shared resources allotted for a given year, stealing a half year from our collective future. Every year. We can only live this way for so long before the bottom falls out. The death of our current way of life—our lifestyles, worldviews, fossil-fuel-powered conveniences—can be overwhelming.

We need spaces to be in this knowing together, to weep, mourn, and rage with other people who also feel this truth in their bones. We must learn to hold the tension of so much loss and our complicity in those losses together. We hold no promises of an unrealistic future, and we refuse to submit in resignation. We create space to honor this time of transition as we work on those seeds that might grow into life-centered paradigms. Fellow Good Griever Catie Gould shared about her time in our GGN spaces:

> We weren't trying to put on a good face and reassure each other things will be okay. None of us believed it would be okay. It was incredibly refreshing to exist in our worst-case scenarios and darkest fears together. I left feeling a much stronger sense of how common these thoughts are and the realization that most people are just hoping for some sort of opening to talk about it. The hours spent discussing these really emotional issues made me more comfortable talking about the climate crisis like any other topic, a practice that I want to continue to grow.[4]

Nothing is easy when it comes to death. When we let it in, it hurts, and it's disorienting. Death can feel cruel or unfair, signaling our bodies to shut down and constrict out of fear. It's almost too much for our minds, souls, and bodies to take. And certainly, the dominant culture's focus on preventing aging does not make it any easier for us to understand or cope with the seasonality of our lives and eventual death. We have a real phobia concerning decay. These aspects of life are cast into the shadow and manifest in more harm being done. (We discuss the shadow more in Step 4. For now, suffice it to say that the shadow is the part of each of us that has been fragmented and cast aside, deemed unworthy in some way.)

The author, professor, and veteran Roy Scranton suggests that the ongoing losses we are facing and will continue to face can be more bearable if we learn how to die.[5] Each morning, as a soldier in the Iraq War, Scranton would "imagine getting blown up by an I.E.D., shot by a sniper, burned to death, run over by a tank, torn apart by dogs, captured and beheaded, and succumbing to dysentery."[6] It was never a pleasant exercise, but he learned that, through this process, he became less afraid of death actually showing up to greet him. This practice, this piece of wisdom, was taught to him by Yamamoto Tsunetomo in *Hagakure: The Book of the Samurai.* Tsunetomo asserts that by meditating on death twice daily, one becomes able to live as if he were already dead, thereby gaining freedom in life. This freedom, borne of letting go of our fear of death, allows us to respond to the current situation, to be alive in the present moment, and to honor our mortality.

Unless we learn to see death as an invitation for deep presence and meaning, we will run from it, hide from it, deny it, or do almost anything else to avoid the pain and discomfort around it. Yet, deep down, we all know the truth: everyone we know and love will die. We, too, will die. Death is a part of life. What would our lives look like if we accepted and even honored death? How would it change our collective response to the climate crisis and the mass extinction event unfolding this very moment? How would embracing death alter our day-to-day lives on this warming planet?

JOURNALING EXERCISE: SEEING GRIEF AS LOVE

Think of a loss you have experienced in life. Is it a person, place, home, animal companion, species? Spend a few minutes writing down all of the qualities of your lost loved one.

- What was their name? What is this person, place, home, animal companion, species known by?
- Bring this entity into your mind in as much detail as possible. Is there a smell or a texture you can recall? Is there color that comes to mind? Is there a sound or a voice that needs to be

remembered? Something else? Take some notes on the sensory details surrounding the one you lost.

- What are some characteristic qualities of this entity? What was the entity like? Can you recall a specific memory of the one you lost?
- How did you interact with your beloved? What was your relationship with them?
- When was your last interaction? What was that like?
- What do you miss about them? What do you long for?
- What is one thing you can do today to honor their memory?

As you enter into the grief of what has been lost in your life, what feelings arise? Are they heavy or painful? Do you notice any pleasant feelings such as love or joy? How do grief and love relate? If this exercise stirred something in you, practice it again for another loss in your life.

The Deadly Search for Immortality

From a very young age, we come to an awareness of our finite presence on this planet. From that moment of awakening and for the rest of our lives, we are constantly managing the terror that awareness brings. Yet somehow, that knowledge, the terror of our mortality, does not dominate our daily lives—at least, not that we are *aware* of. Many psychologists and social scientists believe our unconscious fear of death affects us more than we think.

In his 1973 Pulitzer Prize–winning book, *The Denial of Death*, the anthropologist Ernest Becker posed an interesting theory: an underlying fear of death shapes nearly everything we do. Becker claimed that, like other animals, we humans have a biological instinct to survive. But unlike the rest of the animal kingdom, we have a unique awareness of our mortality. Not only are we self-aware and self-reflexive, but we are also capable of thinking about the future—including the end of our lives. Becker argued that this awareness of our death leads us to have a deep, existential terror called *death anxiety*.[7]

Death anxiety is much more insidious than those morbid thoughts keeping us awake at night. Beyond being self-aware, self-reflexive, and able to imagine the future, we humans are also unique in that we can think symbolically. We can assign meaning to experiences, people, places, or things—and we do not always do this consciously. The result, Becker says, is that we can think symbolically about our mortality and often do so without realizing it. Becker argues that much of the fabric of our culture is woven from our fear of death and how we attempt to escape our mortality by investing in something that will outlive us. In short, "death anxiety drives people to adopt worldviews that protect their self-esteem, worthiness, and sustainability and allow them to believe that they play an important role in a meaningful world."[8] This can play out in ways that promote connection and generation or in ways that increase fracturing and Othering.

Terror management theory (TMT) was inspired by Becker's idea of death anxiety and advanced by studies investigating the many ways death anxiety manifests in individuals and the collective.[9] TMT focuses on how managing our existential fear surrounding our own mortality does not always come out in a literal way; it manifests in symbols and through culture. If we know we are going to die, individually, we start looking for ways to live on, to assert that we exist, that our lives matter. There are many examples of how death anxiety manifests through religion, such as the belief in an afterlife, or through writing the next classic novel that will have people reading our words long after we are gone. Maybe our death anxiety is channeled into joining a movement that will live on beyond our mortal body or skydiving to assert how alive we are. We could even tie our legacy to a lover, a powerful leader, or our finances.

This discharge of death anxiety may not be inherently wrong or harmful, but it does become problematic when we assert our culture or our legacy over others and their safety. Studies have also discovered that if one is not practiced in facing their own mortality and the resultant feelings, thoughts of death, even if just on the fringes of our consciousness, can increase bigotry and Othering. In a death-denying culture, thoughts of our demise can feel threatening to us, moving us out of our window of tolerance and into hyper- or hypoarousal, where we are focused on our survival. Research in TMT suggests that by coming to terms with our own mortality, by bringing death from the shadows of our unconscious into conscious reality, we can respond to death with more care, connection, and reverence for life.[10]

DEATH ANXIETY IN A CLIMATE-CHANGED WORLD

> Power does not corrupt. Fear corrupts . . . perhaps the fear of a loss of power.
>
> —John Steinbeck

In the context of the climate crisis, death anxiety plays out in dangerous ways. Not only are we being pitted against one another but we also tend to defend ourselves and our in-groups with clever mind games, doubling down on our biases, defense mechanisms, and overall worldviews. Our death anxiety can keep us biased and ignorant—we can believe so deeply in our own cultural identity and in the symbols that prop up said culture that facts and logic are of little importance.

Janis Dickinson, professor emerita of natural resources at Cornell University, published a provocative theoretical paper linking TMT to climate change.[11] Dickinson suggests that if we are not practiced in accepting the severity of the predicament, we tend to respond in three key ways as the climate crisis evokes our fear of death. One, we deny that climate change exists. Two, we accept that it exists but deny that humans are the culprits. Three, we downplay the effects of climate change, projecting it far into the future, where it will not pose a direct threat to our well-being. Furthermore, Dickinson has argued that as the impacts of climate change increase, we will likely see the following:

- "People will hold more tightly to their worldviews—religious fundamentalists will become more fundamentalist, people will embrace materialism by buying more stuff, and sustainability advocates will become more strident trying to get others to behave like they do."
- "A fearful public will become more easily manipulated and deluded into a false sense of security or salvation."[12]

Our fear of death is putting us—and a habitable planet—in grave danger. Though penned over a decade ago in 2009, Dickinson's paper predicted the rise of Trumpism in the United States (along with other authoritarian regimes worldwide). Dickinson argued that as the effects of climate chaos increase our exposure to thoughts about death, thereby leading to "blind following and reduction in the rational criticism of public figures, particularly charismatic leaders." Dickinson's theory became a reality as the idolization of Donald Trump gained momentum during a campaign built on racism and antagonism toward out-groups. And the Othering and fearmongering continue in the far right even after Trump left office.

We are seeing a crumbling of the dominant paradigm with death anxiety responses propping it up. No amount of proof or facts will dissuade those who believe in a certain worldview. It is not about the facts. It is about the power that is being asserted. Empty promises and confidence from a strong figure satiate the terrors of the world, moment by moment. They temporarily subdue our survival responses. For many, the legacy tied to the American Dream and "freedom for all" could not be failing; otherwise, those who fully believe in these ideals would need a new worldview. The existential pain, the symbolic death, of confronting a failing worldview is too overwhelming, so it is kept afloat even as it corrupts and destroys. These death-defying mental gymnastics may symbolically extend our lives. In reality, however, our fear of death is jeopardizing a habitable planet and those most vulnerable among us.

As the very real combined threats of the climate crisis and social upheaval increase, the reminders of our impending death are constant and staggering. If we do not have healthy ways of collectively discharging our death anxiety and bringing our awareness of mortality into our conscious life, we can expect to see an increase in bigotry and violence toward marginalized groups and those perceived as different from us.

In recent years, death anxiety has taken form as a rise in ecofascism—wherein fascist ideology is applied to environmentalist views to blame the climate crisis and resource scarcity on marginalized populations. Elements of "racial purity and protecting the homeland," the same ideas that lay the foundation for white nationalistic thinking, have been part of the far-right environmental rhetoric since the conservation era.[13] This thinking has grown worldwide but is especially prevalent in the United States. When political leaders and corporate news media espouse hateful rhetoric toward marginalized populations, focusing blame on them (think "border wall," "caravan of migrants," and "dangerous thugs"), our fear responses and biases can get caught up scapegoating and blaming too. When our safety is perceived to be at risk, it is much easier to finger point at a group of people we might consider different from us than sit with the complexity of our time. While we were editing this book, an eighteen-year-old man murdered ten Black people in a grocery store in Buffalo, New York, after penning "I am simply a white man seeking to protect and serve my community, my people, my culture and my race" online.[14] As our imagined futures are unraveled in the Long Dark, the collective fear will be palpable. We will see more of these violent acts targeting marginalized groups. Those of us called to usher in the heart-centered revolution must commit to acting out of compassion and clarity not fear and blame.

We are all going to die—we have no choice about that. But we do have the power to choose how we cope with our mortality. We can push away our death anxiety, and for a little while that comforts us. But sooner or later, it comes out in powerful and unsettling ways. Or we can choose to be aware of the way our anxiety about death shapes our lives. Only then can we begin to honor death and grow as individuals and as a culture.

EXERCISE: DEATH VISUALIZATION

The Buddhist notion of impermanence teaches us that we will lose everything, including all that we love. We can practice letting death into our conscious reality by entertaining loss, sitting with our emotions, and processing them. Use this exercise to help loosen your grip on all that you know and hold dear. This practice is designed to help you let go of what you are grasping after—the things you think you cannot live without, your ideas and assumptions, your culture and worldview, your very life. As Tsunetomo suggests, a regular practice of meditating on our death can aid in transcending the fear, avoidance, and discomfort that come from trying to deny the inevitability of death. Practice this exercise at least once. If it stirs something in you, consider starting your mornings with it.

Think of three of your favorite things: Imagine packing these things into a box (maybe you need a *big* box for this). Take a moment to express your gratitude for the place these things have in your life. Next, envision yourself bringing your box to the bank of a river. Take a few long, slow, deep breaths and send the box floating down the river, never to return. Take a moment to check in with your body. What sensations or feelings are present? What does it feel like to let go of your favorite things?

Think of your home: Take a moment to feel gratitude for your home. Now imagine your home being torn apart piece by piece until there is nothing left. After seeing the dirt and debris, take a moment to check in with your body. What sensations or feelings are present? What does it feel like to let go of your home?

Think of someone you love: Take a moment to feel gratitude for your loved one. Now imagine the death of this person. See their body lying still, breathless on a bed. Feel their hands; kiss their face. Take a moment to check in with your body. What sensations or feelings are present? What does it feel like to let go of someone you love?

Think of your body decaying: Take a moment to feel gratitude for the body that has cared for you in this journey. Then, envision yourself returning to Earth. See the worms wriggling around your arms and out of your ears. Feel the insects climb across your belly. Notice the cells as they decompose to molecules and the molecules as they become elements in the soil. Take a moment to check in with your body. What sensations or feelings are present? What does it feel like to let go of your body?

Think of your whole life and identity: Bring to mind all of your thoughts and ideas and your identity. See everything you have ever created. Take a moment to feel gratitude for yourself and the life you have lived. Put your whole life as you know it into a box and watch as the box gets picked up and carried away by a tornado. Take a moment to check in with your body. What sensations or feelings are present? What does it feel like to let go of your life and identity?

Think of your culture: Take a moment to feel gratitude for the gifts that your culture has provided. Now imagine your culture as a fifty-story building. A powerful earthquake comes and vibrates your building apart, piece by piece. Watch as your cultural norms, expectations, traditions, and stories fall away until there is nothing left of the building. Take a moment to check in with your body. What sensations or feelings are present? What does it feel like to let go of your culture?

Think of your worldview: Take a moment to feel gratitude for the beliefs, values, and ideology provided by your worldview. Imagine Earth as it rotates around an imaginary axis. Starting at the top of the planet, see your worldview detaching from the planet. It loosens and unravels more with each rotation. When your worldview has been completely separated, watch as Earth shakes the worldview into space, leaving nothing but vast openness. Take a moment to check in with your body. What sensations or feelings are present? What does it feel like to let go of your worldview?

Come back to your body. Come back to the present moment. Observe your breath. Wiggle your fingers and your toes. Take three long-exhale breaths and show yourself some gratitude for your practice today. It takes great courage to let go of everything you know.

HONORING MORTALITY IN THE LONG DARK

We're carving our names so deeply in the tree of life that we look as if we're about to kill it. And that comes from fear of our own mortality.

—Clover Hogan, *Force of Nature* podcast

Over the past few years, we have noticed something that falls in line with Becker's theory: the more chaotic the world is, the more often we see some resistance about this step in our support group circles. Some of our members tell us they do not have a fear of death, but more often they bring up ecocide. "I can understand that I'm going to die," they say. "What I can't deal with is that we're killing so many beings and destroying ecosystems. How can a human-made crisis be honorable?"

It is a valid question. Five mass extinctions have punctuated Earth's history. All those extinctions were set in motion by natural forces, like a massive meteor striking Earth's surface or geologic shifts that changed the climate. Our planet is currently experiencing its sixth major extinction event—and it is the first of its kind. This extinction is the only one caused by a single species: *Homo sapiens*. We have given our species the scientific name meaning "wise man," and yet we are killing off billions of lives and our life-supporting systems.

No inch of this planet is untouched by our actions, whether directly or indirectly. We have destroyed habitats, killed off top predators, and caused our planet to warm faster than it has in two thousand years.[15] Species cannot adapt quickly enough to the profound changes we have made around them. Scientists estimate that the "current rate of species loss varies between 100 and 10,000 times the background extinction rate."[16] When one species is killed off, it triggers secondary extinction events, leading to a cascade of death wherein species after species meets its permanent end. A healthy ecosystem cannot exist without biodiversity. An altered system then degrades neighboring ecosystems. A broken cycle leads to other broken cycles. Everything is interdependent and interconnected.

If you are reading this book, you are not alone in holding the following question: *How can we possibly enjoy life when we are so painfully aware of its fleeting nature?* From forests to shorelines to mountains, the very places we escape to seek solace, beauty, and wisdom are dying—and at our own hands. How can we accept—let alone *honor*—a death that may feel like murder? A murder we have a part in?

Another common thread that may arise for you when contemplating and working with this step is not necessarily a fear of dying but a fear of suffering. Projections suggest that the future may include mass violence, competition, and food insecurity. We may fear having to witness as a loved one suffers or as we, ourselves, suffer. The anxiety of a future world rampant with struggle is enough to shut us down in the present moment.

Being stuck in anticipatory suffering diminishes futures we could be cultivating. This is our survival response throwing us into a state of freeze. adrienne maree brown offers another emergent strategy principle and key Buddhist teaching: "What you pay attention to grows."[17] We can harness our attention and energy for acts of creation. For those of us whose direct survival is not threatened, it is our obligation to imagine and begin cocreating what comes next. We can dream cooperative futures where we work together to preserve what is left of ecosystems, species, and relationships. By bringing our imagined future into conversation with others, we just might be able to build this future instead of spending our mental and emotional energy on projections of hardship.

Of course, there will always be suffering in the world, but we can learn to identify which of our struggles are self-inflicted, which are due to unjust social systems, and which are a part of the human condition. The Buddha helps us discern between them in the *Sallatha Sutta*, the story of two arrows.[18] In this story, a monk is struck by an arrow, which causes pain. The untrained monk grasps at the arrow in a panic. His reaction to the first wound doubles his pain—and that is the metaphorical second arrow. The Buddha explains that resisting or overreacting to the first arrow causes more pain than the initial injury. If we can learn to accept the initial pain rather than try to avoid it or tell stories around the first arrow, we can lessen our suffering surrounding the second arrow.

The first arrow is inevitable in this life. Events like illnesses, accidents, injuries, and death happen. But we can learn to avoid the sting of the second arrow. The self-infliction of suffering over our projections of a future world that has not yet come prevents seeds from being sewn. Worrying over something that has not arrived robs us of our present moment and our opportunity to imagine a collaborative future world. Our energy is best placed in dealing with first arrows and dispersing seeds.

EXERCISE: CREATING YOUR PERSONALIZED LIST OF RESOURCES

What and who helps you feel grounded and safe? These are your resources. It's important to identify them so that when you are thrown out of your window of tolerance, you can call on your resources to help you come back to the present moment and reestablish a sense of okayness. Using your unique set of resources may help you restore regulation in times of distress.

Using the questions below, make a list of your resources. Keep your list accessible so that you can come back to it time and time again.

- What is a breathing exercise that helps soothe you? How can you use your breath to come back to the present moment and feel into your body?
- How can you move your body to help ground you? Do large, macrobody movements like shaking help calm you? Or do you prefer small, micromovements like toe wiggles?
- Do you have a quote, lyric, phrase, or mantra that calms or strengthens you? Maybe reciting a beloved poem by heart is a resource for you.
- What is a grounding power object that you can make physical contact with to help you come back to yourself? Perhaps it's a crystal, stone, necklace, ring, or beads.
- Who is a living being who can support you or has supported you? Does a family member, friend, or animal companion come

to mind? Perhaps your resource is a tree or an animal you do not know but find a sense of strength or solace through envisioning them.

- Who are your ancestors, deities, or gods who can serve as guides along your path?
- Which places help you feel calm and centered? Is there a place you go to that you can transport yourself to through your imagination? Perhaps you feel deeply connected to a particular beach or forest. Maybe you feel the most centered and calm in your bed or at your kitchen table.
- Is there something else that is coming to mind that ought to be added to your list of resources?

We said it in Step 1, and we will say it again here: acceptance of the predicament does not mean giving up on working toward new paradigms. Likewise, honoring death does not mean throwing our hands up and letting this sick culture destroy our planet. The word *honor* comes from old Anglo-French, meaning "to revere or pay respects." This step in our program used to be titled "Confront My Mortality and the Mortality of All" until we had a conversation with our friend and colleague Michael Dowd. Michael reminded us that to confront is to have an aggressive encounter, but to honor is to embrace, to see death as part of life, and to build a relationship with our mortality—to accept our first arrows with grace and become aware of when the second arrows are stealing our energy.

To honor mortality, we must acknowledge the ultimate power of death—and if we acknowledge death's power, we can no longer hide from our existential anxiety by repressing it. Honoring death means practicing being with deeply uncomfortable feelings and making peace with the ultimate uncertainty so that we can make sure dying is not getting in the way of living. In doing so, we can harness our fear of death and transform it into a fire for life. Ask yourself: *While I am alive, what kind of life do I want to live? If I am going to leave my footprint on this planet—and, inevitably, I will—how can I make my legacy a positive one? What kind of ancestor do I want to be?* When we honor our mortality, we can ask these questions of ourselves now instead of just before dying, when it is too late.

Both Life and Death

> Live to the point of tears.
>
> —Albert Camus

Living an openhearted life in the Long Dark is no easy task. Being fully alive during times of chaos when we are confronted with so much death, requires that we develop a special emotional skill: *both/and thinking*. It connects to the ancient Chinese concept of yin and yang, light and dark, life and death. Both/and thinking removes binaries and instead creates space for complexity. Even if that complexity seems messy. But before we can start honing our both/and practice, we have to recognize and avoid either/or thinking.

In psychology circles, either/or thinking is often called all-or-nothing thinking or splitting. When we engage in either/or thinking, we see the world in binaries: black or white, good or evil, sad or happy, natural or unnatural. Either/or thinking is considered to be a defense mechanism, a polarized way of seeing the world that helps our brains act quickly under stress. There are times when either/or thinking helps us survive. For example, when our ancestors saw a saber-toothed tiger on the prowl, their brain immediately categorized the situation as "bad." They likely did not stop to think about the more accurate, less polarized truth: "A saber-toothed tiger may be bad for me right now but good overall, because they play an integral role in maintaining the biodiversity of a balanced ecosystem." Now that is an excellent example of both/and thinking. We may well have had ancestors who stopped to ponder both the pros and cons of predation, but if they did exist, they did not last long enough to tell the story.

Either/or thinking is rooted in life-and-death situations, so it is easy to see how this mindset could be triggered by death anxiety in the present day. Even though this way of thinking can help us survive, it can also be a cognitive distortion, or a pattern of thinking that is irrational or blown out of proportion. Cognitive distortions are often associated with depression and anxiety. For many, modern-day living has guarded us from daily life-and-death situations, but still, either/or thinking persists.

While either/or thinking has value, especially in potentially dangerous moments, it also comes with incredible limitations. Much like many of the defenses we put up against death anxiety, either/or thinking restricts the way we see the world. It reduces complexity and forces a simplified perspective, which is not helpful in times of unraveling or when cocreating new ways of being. This way of thinking perpetuates a dualistic orientation, making us believe that if I am right, someone opposing my ideas or beliefs must be wrong. Either/or thinking might sound like these examples shared in our GGN circles:

- "Humans are causing climate change and we're killing everything. There is nothing I, as one person, can do about it."
- "It's so hard to talk to my mom. She always argues with me about the climate crisis and gets her 'facts' from Fox News every day. They fill people's heads with lies and propaganda. I wish she'd listen to me when I tell her that."
- "The climate crisis is only going to get worse, and my kids are in for a life of suffering. I am a bad parent for even having them."

With some practice, we are capable of holding the tension and paradox of two seemingly opposing truths, which is what both/and asks of us. Here are some examples of holding the complexity of the collective moment we are in:

- We are losing so much, and there is still so much beauty around me.
- The amount of suffering in the world is excruciating, and I will fully embody my life and be present for the full range of my life experiences.
- I am often deeply sad about the state of the world, and I am pursuing joy in my daily life.

It takes effort and practice to pull ourselves out of either/or thinking and reorient our brains toward both/and thinking. We can start by learning to recognize black-and-white thoughts when they come up. Some red flags are words like *never, always, every, completely, should, impossible, evil, bad,* and *but,* just to name a few. It is important not to shame ourselves for thinking either/or thoughts—after all, depending on the situation, they can be both harmful and helpful (see what we did there?). Becoming more aware of these thoughts will help us start to notice when they show up, and we can begin practicing both/and thinking.

Once we have identified an either/or thought, we can try looking at it from a both/and standpoint. Here are a few examples of both/and thoughts. Some helpful words for both/and thinking are *also, at the same time, sometimes, maybe, might, could,* and of course *both* and *and.*

EITHER/OR	BOTH/AND
Humans are causing climate change, and we're killing everything. There is nothing I, as one person, can do about it.	My individual actions might not make a big enough change to prevent the worst impacts of the climate crisis, but I feel that my actions should not be outcome dependent. I will do what I can because it is in line with my values.

EITHER/OR	BOTH/AND
It's so hard to talk to my mom. She always argues with me about the climate crisis and gets her "facts" from Fox News every day. They fill people's heads with lies and propaganda. I wish she'd listen to me when I tell her that.	My mom and I have a complicated relationship and do not see eye to eye about the severity of the climate crisis. I can be frustrated with my mother and show her compassion.
The climate crisis is only going to get worse, and my kids are in for a life of suffering. I am a bad parent for even having them.	I am sad that my children will be more affected by climate chaos than my generation, but I also see that I can teach them to embrace life, joy, and how to build resilience in times of great suffering.

Both/and thinking paints a richer—and more accurate—picture of the world we live in, and it can make us more compassionate toward ourselves and others. Earth is both in danger and full of life and meaning, and we are both the stewards and the destroyers of this planet. We must learn to balance both grief and joy to live our all-too-brief lives to the fullest.

EXERCISE: INTUITIVELY MOVING THROUGH GRIEF

Moving our bodies is a key part of processing our emotions, particularly grief. We see this in cultures where dance, processions, and other body movements are integrated into mourning and burial rituals. Similarly, this exercise uses body movement to embody and process grief.

Create a playlist of songs that remind you of the people, places, and/or things you have lost—a loved one, an animal companion, a beloved place, a home, a land feature, a particular time in your life, or a species. Then, find a space where you can open to fully feeling, vocalizing, and moving through your grief. If you are feeling self-conscious about intuitive movement, it can help to find a private space and cover any mirrors that might distract you from the purpose of the exercise.

There is no right way to move through grief. Listen deeply to your inner world and see how your soma wants to move. Your movement may be upbeat or slow, or it may oscillate between both. It could resemble a practice like yin yoga, with long, slow stretches. You can even lie down on the floor, pausing in stillness to feel where you can surrender to the gravitational force of Earth. Notice any feelings that arise. You may find that they are seemingly contradictory. Ask the feelings if they want you to move in any way. Welcome any sounds your soma wants to release. Maybe it's a yowl, a scream, or a roar. Perhaps you need an uninhibited wail. Open to any tears that need to flow. As you move grief, you may even feel joy in your body. There is space for all feelings.

As you end your intuitive movement practice, take a few minutes to reflect and integrate the experience. Did anything surprise you about this practice? What types of feelings and sensations were present at the beginning, middle, and end? Do you feel differently after moving than before you started? If you liked this practice, set a date with yourself to do it again. Consider inviting a few loved ones to join you for an intuitive grief movement party. Choose new songs or try the same playlist again and see how your experiences compare.

I Am That

LaUra

The ancient phrase *tat tvam asi* has many translations; my favorite is "I am that." The quote is originally from the Hindu text *Chandogya*

Upanishad and is a constant reminder to keep my heart open to my inherent connection to all that is, even if that connection feels painful sometimes. I am a small part of this vast and dynamic process of life, death, love, and reciprocity.

On an autumn day, I am flat against the floor in front of the open sliding glass door of my one-room apartment. The sun shines in and warms my skin. A cool breeze carries different scents: decaying leaves, car exhaust, a neighbor's cigarette smoke, and the crisp, dry air that warns winter is near. Despite the deep sorrow flowing through my body, each touch of the wind, the variety of scents, and the warmth of the rays on my skin all remind me that as death permeates this moment in geologic time, this planet is still very much alive and in transition. I am very much alive and in transition. As my heart beats, stomach gurgles, and throat tightens from being with my sadness, I tell myself that the immense grief I feel over the state of the world is a direct reflection of my love for the world. I lean into this complexity—I live in the in-between and practice both/and thinking. *Tat tvam asi,* I am that.

All life is valuable because of the interconnection and interrelatedness existing throughout the entire web of life. The pain and loss that the world is undergoing is impossible to unsee. I think of my niece who taught me at age five so much about interbeing. "Animals being hurt makes me sad," she once told me, "because they are special and part of this world." We can make daily choices to protect fellow Earthlings and to ease suffering. Learning how many species are being forced into extinction, and then understanding the immeasurable number of individual lives lost within those species, requires similar tools as those we implement for grieving the loss of a friend. We must lean into the discomforting feelings, create space to process them, move our bodies so the feelings do not get stuck, and reduce suffering through mindful actions. Though, collective griefwork cannot be a solitary act. The magnitude of loss is too much for an individual and we are communal creatures. We must remember and create ways to move collective grief in community.

EXERCISE: RITUAL FOR THOSE LOST[19]

Choose a time and place where you will be able to fully and openly feel. Gather some pebbles or twigs. Light a candle and play some music that perhaps has special meaning to you. Write a list of the names and things you have lost and want to honor (a place, ecosystem, person, animal, entire species, etc.).

One by one, hold up a pebble or twig to represent the loss and speak their name aloud. Speak to the pebble or twig, honoring one thing about that being. For example: "Honeybees. I love how honeybees help my garden flourish." Set the pebble or twig into a bowl or piece of cloth. Do this same practice for the next loss on your list and the next, until your list has been read and honored aloud. Consider reading a poem about loss or grief.

Bring the pebbles or twigs to a river, stream, or other body of water and release them into the water, saying, "Today, I honor my losses and return them to Earth."

Repeat this practice regularly as a way to invite in and honor the losses in your life. Lead your community through this exercise and see how the experience changes. Grief is best moved in community.

Step 4

DO INNER WORK

> If you begin to understand what you are without trying to change it, then what you are undergoes a transformation.
>
> —Krishnamurti

When we think of solutions to climate breakdown, we often jump to organizing protests, policy changes, and big technosolutions like windmills, solar power, and electric cars. These solutions are important in our transition away from fossil fuels—and they are largely incomplete. The overlapping crises we are experiencing are much deeper than the climate crisis alone. To focus on policy and infrastructure without serious inner work will have us repeating the same issues that got us into this mess. If we really want to create paradigms that protect and preserve life, we have to dig deep and understand that these problems did not occur by accident. The global predicament is centuries (nay, millennia) in the making because of severely flawed worldviews based on exploitation, Othering, and disconnection. To undo these worldviews, we must turn inward and do our inner work.

But let us be clear: when we say inner work, we are not talking about self-improvement bullshit that says you need to "fix" yourself. You are not broken. No—not even if you spend days in bed with depression or lash out at a loved one when you are under stress or choose not to do whatever that thing is that you know is good for you but for some reason you just cannot bring yourself to do it. These things do not mean you need to be fixed. They simply mean that you, like everyone, have some work to do. As we have mentioned many times throughout this book, we all must feel our feelings and heal our traumas. Susanne Moser suggests that it is important to examine our baggage and learn how we become trapped by our past wounding. If we do not deal with the issues, unprocessed feelings, and traumas in our personal lives, we will bring them into current and future struggles.[1] Because we invite trauma healing elsewhere in these pages, we focus now on integrating the shadow aspects of ourselves—an unconscious part of our psyche that most of us would rather not look at and that we wish could be hidden, tamed, or banished with self-help books.

Into the Shadows

> I would like to play the part of someone who has worked on my consciousness sufficiently so that if things get tough, in terms of environment, social structures, oppression, minority groups, whatever the thing is—I would like to be able to be in the scene without getting caught in my own reactivity to it, without getting so caught in my own fear that I become part of the problem instead of part of the solution.
>
> —Ram Dass, from his lesson "For Those Attached to How Things Were"

The concept of a shadow originates in the work of the late psychoanalyst Carl Jung, whose philosophies of the human psyche forever changed the way psychologists think about consciousness and the self. According to Jung, learning to embrace our shadow self is critical to becoming the best version of ourselves. He wrote volumes about consciousness and the shadow. But thankfully, you do not need to be an esteemed scholar of Jungian psychoanalysis to start doing shadow work. You can start right now.

Here is the simplified version of Jung's philosophy of the shadow self: Jung believed that a person's psyche is composed of a few different parts. For the purpose of this step, we focus on just two: the ego, or the conscious part of ourselves, and the shadow, a part of our unconscious self. The ego is any aspect of ourselves we are aware of and identify with. Our job title, ethnicity, regional identity, gender, and personal histories are parts of our ego, to name just a few. In the dominant paradigm, we often use the word *ego* with a negative connotation—"That guy has a huge ego!" But in Jungian theory, the ego is neither inherently good nor bad. It is, however, pretty self-centered. Because the ego is only what we are aware of, the ego sees itself as the *entire* self. Someone who operates solely from a place of ego has no idea that their actions, personality, and feelings are also being affected by their unconscious.

Now, the shadow self, despite its name, is not necessarily our "dark side." Rather, the shadow self is kept in the unconscious—it lives among our dreams, desires, emotions, impulses, experiences, and the best and worst of our personality traits, all of which are hidden from our conscious selves. The shadow is constructed of the fragmented parts of ourselves we often learn to hide or repress from a young age not because they are inherently bad but because we *perceive* them to be bad. We tend to repress pain from traumatic events and hide the parts of ourselves that we have been taught are socially unacceptable or shameful. But oftentimes, our shadow also hides our personal power.

MY PERSONAL SHADOW

> One does not become enlightened by imagining figures of light, but by making the darkness conscious.
>
> —C. G. Jung

Aimee

I was twenty-one years old the first time I checked into a psychiatric ward. It was my junior year of college. I had been struggling with depression for some time and—though LaUra was the only person who knew it—I had started engaging in self-harm. It was the middle of the semester, and we should have been pulling all-nighters studying for midterms. But instead, LaUra stayed up all night watching over me as I had a total mental breakdown. When dawn broke, she refused to let me go back to my apartment.

"If I let you go back, I worry you might try to . . . kill yourself," she said. "And I can't let you do that."

"You're overreacting," I told her. "I'm fine."

I had repeated this lie so many times in my life that I actually believed it. In high school, my life looked perfect from the outside: I got great grades, I was a dedicated athlete, I helped out in the community. But LaUra was not buying it. She gave me a choice: I could go to the hospital in an ambulance, or I could go to the hospital with her. Either way, I was going.

I ended up taking the trip with LaUra from our college town to a small, rural hospital—not once but three times, all in just two months. My first two hospital stays were short. I would stay for a long weekend, dutifully attending individual and group therapies every day. I was an easy patient, easier than they were used to, polite. I smiled a lot. "I'm doing OK. Thank you for asking," I would say, and after a few days, they would let me go.

On my third stay in the hospital, I found something that would change my life: a rogue paper clip in the hallway. This was tantalizing illegal contraband, a weapon of mass self-destruction. Someone could kill themselves with a paper clip, a nurse once told me. When no one was looking, I picked it up, stashed it in my sleeve, and decided to try it out for myself.

Later that night, after lights out, I snuck the paper clip out of my waistband and gouged it into my wrist. I dug the dull metal into my skin again and again, doing the most damage I could—which turned out to be not much. The suicide attempt was futile, but I had gotten away with it.

The next day was to be the day the doctors cleared me for release. But first, I had a group activity to attend. The staff gave us all coloring books and soft-tipped magic markers. As I colored neatly inside the lines, I waited for the therapist to notice the scrapes on my wrist, but no one did. At the end of the session, I lingered in the doorway, making small talk with the therapist. And then, for reasons I could not explain at the time, I confessed. I didn't have to, and if I hadn't, I surely would have been back in the comfort of my apartment within a few hours. Instead, a group of nurses and doctors swooped down on me, ushering me back to my room and strip-searching me to ensure I hadn't hidden any other potentially harmful objects on my person.

Why had I done something so pathetic, I wondered, *and why did I have to go and tell someone*? At the time, I didn't understand why I had done what I did—just that I felt mortified. But in retrospect, it is clear: the part of me I had denied so long—the depression, anger, and shame that I had disconnected from and cast away—needed to be seen. Thomas Moore, in *Care of the Soul*, explained that the soul acts out when its deepest needs are not being heard.[1] My shadow was demanding attention and healing. It knew better than I did how to help me.

To this day, being strip-searched over a paper clip is still one of the most humiliating things I have experienced. But it also saved my life. It was my first step toward embracing the aspect of my shadow self that I was most ashamed of: the depth of my emotion. All my life it seemed like I felt sadness, passion, and joy much more strongly than my family and friends. They were especially unreceptive to my passion for activism. The message I received from those around me was clear: my strong emotions were abnormal and socially unacceptable.

We all have a shadow side, no matter how hard we try to hide, deny, or heal it. But the shadow self does not need fixing. It needs to be embraced—or, as Jung says, integrated. Jung believed that when we remain unaware of our shadow, our personality remains fragmented.

These unexpressed fragments control our behaviors, relationships, and moods—unbeknownst to us. For me, my repressed anger, sadness, and shame manifested as self-injury and depression.

But over the years—and with loads of therapy, support from loved ones, and gritty inner work—I have slowly started to embrace my strong emotions. It's an ongoing practice. Some days, my intense feelings make getting out of bed an impossible task. Other days—when I'm teaching intuitive body movement, facilitating a support group, or having a heart-to-heart with a friend—my ability to feel deeply is my superpower. My depth of emotion can be overwhelming at times, but it's also what empowers me to connect with others. That's the beautiful and multifaceted nature of the shadow self: you cannot access the full depth of it without embracing all of it.

What does it mean to truly accept our shadow? It means feeling our full range of feelings—even the ones that make us uncomfortable. It means examining what we believe are the worst aspects of ourselves—even the ones we are ashamed of. And instead of shaming ourselves over what we perceive to be our worst qualities, we work to understand the unconscious motives behind our "negative" traits. We put the word *negative* in quotation marks because no quality is inherently good or bad—no, not even anxiety, lying, disconnecting from our feelings, lashing out at loved ones, or substance abuse. At some point in our lives, they helped us cope with difficulty and traumatic experiences. We can embrace these parts of our shadow by thanking them for their service and then gently breaking the news that we don't need them to survive anymore. If we want to change our perceived shortcomings and mistakes, we have to approach them with compassionate curiosity, so we can begin to understand and even feel grateful for our shadow.

Embracing the fragments of ourselves that we have cast into the shadow is key to doing inner work. And it is revolutionary work because it is the opposite of everything that got us into this collective crisis in the first place.

EXERCISE: WRITE A LETTER TO YOUR PAST SELF

Think back on a time when you acted out of alignment with your values. What were you going through at the time? How might your unresolved traumas have influenced your actions? What were the stressors in your life? What was your shadow trying to bring attention to when you behaved this way? Grab your journal and write a letter to yourself reflecting on as many aspects of the past situation as you can remember. Be gentle with yourself, like you would with a young person or teenager who is still learning and growing. Remember that your past self was doing the best they could at the time with the tools and resources they had. After you finish the letter, take three long-exhale breaths, shake your body out, and consider burning or ripping up the letter, saying to yourself, "I did the best I could at that moment in time, and I am ready to let go of that situation."

Bonus tip: Sometimes, approaching your shadow self with some humor can help ease the pain. We have a friend named Ashley, who had a nickname for her past self: Past-shley. Anytime Ashley talked about Past-shley's mistakes, it was always with a tongue-in-cheek attitude that made talking about her shortcomings a lot easier—and with a bit of humor. Coming up with a playful nickname for your past self might help you look at your shadow aspects in a more lighthearted way.

Healing Our Radical Wound

If you are that alienated from where you live and who you live among, if you are so completely persuaded by this idea that growth is the same

> thing as accumulation and vice versa, then you will never be well. The only well-being is interbeing. What seems alien to us is in fact kin. If we can turn our fear of the alien into a love of kinship, we can be well again.
>
> — Richard Powers, "The Space Between," an interview in *Atmos* magazine

How we treat ourselves manifests in how we treat others—and historically, we humans in the dominant paradigm have done all types of terrible things to each other. In Step 3, we looked at this from the lens of terror management theory: in an attempt to "escape" death, we Other those who threaten us or our worldviews. This notion is present in Jungian psychology, too. Jung called this phenomenon *projection*—that is, the idea that we unconsciously displace our feelings, fears, desires, and shadow aspects onto other people. When we deny our shadow selves, there are collective consequences on a much larger and more disastrous scale. Scapegoating, polarized political parties, and war are all a result of our collective shadows left unchecked.

Similarly, we project our shadow onto the more-than-human world, coloring animals, insects, microbes, plants, weather, forests, mountains, and rivers as threatening outside forces that are separate from us. Think back to the definition of the ego: any aspect of ourselves we are aware of and *identify with*. Those of us living in the dominant paradigm no longer identify as part of the natural world. So many of us do not regard the more-than-human world as a direct part of our daily existence. In *Hospicing Modernity*, Vanessa Machado de Oliveira argues that the "deeper, older violence is the imposed sense of separation between ourselves and the dynamic living land-metabolism that is the planet and beyond."[3] This perceived separation is the *radical wound* from which humans in the dominant culture suffer.

A degeneration process occurring over thousands of years began when we sought to prove how civilized we were. By taming, curating, and controlling our environments to maximize our own survival, we created an illusion that we still live in—we are not animals; we are god-like. We doubled down on disconnection after the so-called Enlightenment period. By disregarding other ways of knowing and putting rational thought on a pedestal, we drove a wedge between our living, dynamic somas and the living, dynamic planet to which we are intimately bound. By prioritizing our minds over our bodies and accumulation of material over relationships, we created more separation, more Othering, more alienation. In this worldview, we exist because we think, not because we feel, breathe the oxygen plants produce, and relate to one another.

Along the way, we have barricaded ourselves in densely packed urban fortresses that have pushed out most things wild. Eighty-three percent of people in the United States live in urban areas.[4] Nature is something we watch whiz by us outside our car windows or on television docuseries. While urban areas can offer more opportunities for community connection and education, many of us have lost touch with the intimate relationship that our ancestors once had with the more-than-human world. We have forgotten that we are, indeed, nature.

By distancing ourselves so far from the more-than-human world, many of us have come to fear it. We subdue our fear with our acts of superiority and supremacy. Our perceived human supremacy gives us permission to dominate and pillage, to build more factories that pollute our air, water, and land, to annihilate entire species, to grow our food with poisons, and to exhaust our planet's resources without much thought. The mountains that were once believed to be the home of the gods are now something to conquer, not by climbing them with oxygen tanks but by blasting them for interstates or mining them for coal. As any climate activist knows, ending these practices is proving to be hard work—and it is just the beginning. Doing the collective inner work to end the unconscious mindset that got us there is a whole new ball game—a critical one—and we are only in the first inning. As more and more of us question *why* we treat our planet the way we do, we can begin to bring these parts of our society's shadow to our collective awareness.

A critical symptom of our radical wound is that we have begun to fear the wild within ourselves. To change our destructive habits, we must venture into shadowy regions of our own somas and embrace the untamed parts of ourselves we find there. Healing starts by rewilding ourselves.

REWILDING OURSELVES

> To be whole. To be complete. Wildness reminds us what it means to be human, what we are connected to rather than what we are separate from.
>
> —Terry Tempest Williams

The climate catastrophe, fragmentation, and biodiversity loss have destroyed much of our ecosystems, and there is a movement—called *regeneration*—to help repair, restore, rebuild, and rewild these landscapes. This is a deeply worthwhile endeavor, and one we believe in with our whole hearts. More important than healing what we broke, we must stop destroying our living, breathing life-support systems that still exist. To do this, we need fresh worldviews that remove our ingrained ideas of supremacy and replace them with values that uphold how intimately connected we are to our planet and the systems that provide clean air, water, and food and a livable biosphere. We must learn to rewild ourselves and integrate our living spaces with the wild world around us.

We have become domesticated, silencing our inner wisdom and desires to conform to the dominant culture. We look to cultural norms and expectations for guidance on to how to be a human being. Cultural messaging through advertisements, home makeover shows, and lifestyle listicles tell us to grasp after things and money and facts. We are taught that happiness lies in the accumulation of wealth, so we can buy more stuff. We turn to scientific studies to tell us whether our pets really are capable of loving us, how much salary we need to feel emotionally fulfilled, and how we can optimize our minds for productivity with egg timers and scheduled naps. We are distracted into conforming, as we move further away from our innate ways of knowing.

Rewilding ourselves means questioning the ways our culture's domestication plays out in our day-to-day lives. From childhood, the dominant culture teaches us to repress and deny many natural aspects of our animal nature and turn away from our embodied wisdom. We receive messages that we are expected to be polite and sit still. We learn to be quiet and "control" ourselves. As soon as we hit puberty, we are told to slather on deodorants and body sprays, contain our breasts in bras, cover our blemishes with creams and face washes, and be ashamed of our body hair. We keep discussions of sex hidden, manifesting in unhealthy expectations and compulsions concerning sex and relationships. As we age, we are targeted for products and procedures to remove wrinkles and look younger. It is no wonder so many of us live disconnected from our bodies—we are constantly told by the advertising industry how wrong our bodies are. And, if that were not enough, so many of us spend most of our days staring at screens, further disconnecting from our bodies.

This is not what our animal somas want. Our innermost yearnings desire more play, imagination, howling, dancing, singing, rest, and listening for the chirps of crickets. We long to engage with our senses and to get our hands and feet dirty, to talk to trees, be near water, and connect with each other in deeply profound ways.

The work of rediscovering the wild within forces us to question the stories we have taken for granted and to become open to new stories about what it means to be a human being alive in these times. Our wild selves often hold innate wisdom that our domesticated selves have long forgotten. But if we can manage to look deeper than our egos or the cultural stories we have been socialized with and really listen to our deprogrammed somas and intuition, we can tap into lost knowledge we did not even know we had. We can begin to see ourselves as deeply connected to and an extension of the more-than-human world. As the author, podcast host, and fitness expert Aubrey Marcus reminds us,

> You are comprised of: 84 minerals, 23 Elements, and 8 gallons of water spread across 38 trillion cells. You have been built up from nothing by the spare parts of the Earth you have consumed, according to a set of instructions hidden in a double helix and small enough to be carried by a sperm. You are recycled butterflies, plants, rocks, streams, firewood, wolf fur, and shark teeth, broken down to their smallest parts and rebuilt into our planet's most complex living thing. You are not living on Earth. You are Earth.[5]

Domestication, by its very nature, is reductionistic, fragmenting complexity into smaller and smaller bits in hopes of understanding, studying, and analyzing it. As hard as we try, we will never have the whole picture by reducing systems in a piecemeal way because, as we have learned from systems theory, the magic happens in the relationships between parts. Life by its very nature tends toward wholeness, complexity, and diversity. By living in the dominant paradigm, we have become fragmented and reductionistic, as individuals. The collective is created out of individuals; by rewilding ourselves, we help rewild the collective. Our deeply held beliefs, values, and ways of being must transition our focus to the principles of life (wholeness, complexity, and diversity).

Where and how we live in the dominant paradigm have contributed to our disconnection from our animal nature. Historically, our urban spaces have been designed to keep the wild out while maximizing human functionality and lowering costs. This way of organizing human societies is not working for us. It affects all of our somas through exposure to a variety of pollutants, constant overstimulation, and the illusion of our separateness from the more-than-human world. Studies show that children who live in urban areas with limited access to the natural world are more likely to be nervous and depressed as adults.[6] Urban kids are also more likely to be exposed to pollutants, and the effects are not just physical. Pollutants can cause cognitive delays and even psychotic breaks in adolescents, which can lead to a higher likelihood of psychosis as they grow into adulthood.[7]

The importance of maintaining relationships with our inner and outer wild worlds is not a new concept to some cultures around the world. The failures of the dominant paradigm are summoning us to overcome the consensus reality, remember our connections, and break out of our concrete jungles. We must learn how to share resources and space with the more-than-human world. An emerging interdisciplinary field called urban ecology is growing in popularity and is helping us reimagine our human landscapes in relation to the natural world. This field "aims to understand how human and ecological processes can coexist in human-dominated systems and help societies with their efforts to become more sustainable."[8] While still viewed through a very human-centric lens, it is an appeal from the dominant culture to start incorporating the wild world back into our isolated, insular environments. In the new paradigms we are creating, we must learn how to be with the wild forces within and outside of ourselves.

As urban ecologists study, map out, and design our cities, we can help our rewilding process along by spending time out of doors as we are able. This can help us remember our animal nature and provide a plethora of health benefits, from lowering blood pressure to increasing our happiness.[9] Just ten minutes spent walking through a natural environment can lessen our mental and physical stress.[10] Time outdoors also helps reduce rumination, or stewing in our spiraling thoughts or the qualities we dislike about ourselves (aspects of our shadow).[11] However, not everyone is able to get their feet in the soil, walk along a riverbank, or saunter through a park. If you do not have access to green space, it is critical to bring wildness to you. Begin growing a plant, bond with a dog or cat, or, at the very least, take in images of the natural world. These actions also help calm our minds and bring us into greater connectivity.

Reconnecting with the more-than-human world is a critical part of stopping the harm we are perpetrating onto our living systems. As an added bonus, reconnecting with our living planet also helps us be more emotionally resilient. We must move from thinking of ourselves as separate from the natural world to feeling our inherent connectedness with Earth.

EXERCISE: ANCHORING IN WILDNESS

Carve out three to fifteen minutes of quiet time during which you are not likely to be interrupted. It is best to do this outdoors and in a relatively quiet natural space, like a park, a forest, or near a body of water. If you do not have easy access to an open space, try to find a window or bench where you can look up at the sky or notice a tree. Wearing earplugs or headphones with ambient music can help drown out city noises.

Notice the trees, plants, birds, insects, or sky around you. Choose something—an object or being—to be an anchor for this exercise. Begin observing specific aspects about the anchor. What are the textures, colors, scents, and sounds of that object or being? Notice if your thoughts start to wander. Are they leaping forward to future thoughts? Are they focused on a past event? If this is happening, that is OK. Notice this, and gently bring your mind back to your anchor. What can you notice about your anchor that you have not already taken in? Is there a new smell, or did the color change? Has it moved? Have your thoughts wandered again? Practice bringing them back to the outdoor anchor.

When you are done, take a moment to reflect on this exercise. Is there something that focusing on your anchor taught you? What did you notice? How did this exercise make you feel?

Repeat this exercise often to help gain clarity and quiet your mind.

DANCE IT OUT

You were once wild here. Don't let them tame you.

—Isadora Duncan

Aimee

In the early days of GGN, a member of our circle, Dick Meyer, offered us a place to live rent free so that we could focus on expanding our program. But there was a catch. We would have to move over five hundred miles away from our home in Salt Lake City to a town in western Nebraska we had never visited or even heard of. The opportunity to build GGN was too good to pass up for us as idealists. So we quit our day jobs, said goodbye to our stable incomes, health insurance, and in-person community, and left our life in Salt Lake City behind. After a long day's drive, we arrived in Nebraska with our minivan, two cats, and two dogs. We had no idea what our new home had in store for us, but one thing was clear: we were officially in the middle of nowhere.

It wasn't long after the move before our 10-Step program started to take off on a new level. On paper this sounds great, but there was a problem: we were not making any money. In case you couldn't tell by now, navigating capitalism doesn't come easy to us. We developed this program to undo our cultural conditioning, not profit from it. We soon found that while we were pouring all our time and energy into our program, monetizing it was a personal and ethical struggle. Meanwhile, we had few local friends to offer us emotional support, and many of the people we met just didn't understand what we were building. As the months passed, and autumn turned to winter, we learned that this region experiences a unique type of winter. This region of the country is what the locals call "Wyobraska." Known for whipping winds, waist-high snow drifts, and negative temperatures for days, the winters prevented us from getting outside and seeking solace in the natural world.

In short, the struggle was real and soon my depression crept in. Not having access to health insurance or money for a therapist, I felt stuck.

Most days, I just wanted to stay in bed and cry—and a lot of days I did. With no funds to join a gym and no yoga community available, I was limited in my exercise options. But amid this horrible season in my life, I had a sudden and surprising urge welling up: I wanted to start *dancing*.

Dancing? This was something I had not done since my party days in college. I had always liked dancing when I had had enough to drink, but the thought of trying it sober sounded intimidating at best and humiliating at worst. But after a few days, the urge hadn't gone away. So I locked the doors to the bedroom, covered up the mirrors, put Kesha on blast, and, just for a few minutes, danced. The next day, I danced a little longer, and the next day, and so on. In all my sadness and frustration, dancing felt like the only thing that made sense. It felt like freedom. I shook, jumped, and stomped. I did whatever my body wanted me to do and let go of all self-consciousness. With a clear mind after moving my body in this way, regulating my emotions became a little easier. I felt strong. I became healthier and established a dance practice—eighty minutes, at least three times a week.

As it turns out, my sudden urge to dance during hard times was ancient, animal wisdom. In the wild, animals move their bodies to metabolize the stress hormones and access energy left over from a stressful or threatening event. Wild animals undergo threatening events all the time—especially prey animals who are always one misstep away from becoming lunch. After an animal narrowly escapes being killed, they start shaking, sometimes so hard it looks like it is convulsing. When scientists studied this phenomenon, they discovered that these tremors had a purpose. When an animal is in a chase, they go into survival mode, and the sympathetic nervous system kicks in, flooding the body with a complex mixture of hormones and energy that trigger the fight, flight, or freeze response.[12] After the chase ends, the animal shakes, which helps calm the sympathetic nervous system and dissipate stress hormones, bringing the body back into balance. Without this shaking movement (or another type of embodied release), the body gets stuck in survival mode, unable to process the stress hormones and return the nervous system to its natural resting state.

The sympathetic nervous system is something we have in common with animals, and though our threats may be different, our body's responses are the same. Getting laid off, losing a loved one, being sexually assaulted, experiencing a flash flood, escaping a wildfire—all these can cause us to have a physiological survival response just like the animal being chased by a predator. And like other animals, after undergoing a stressful or threatening event, our bodies innately know to move, to release. We bounce our knees when we are nervous; our lips quiver when we cry; we shake when we are afraid. Wailing, kicking, screaming—all the things we teach our toddlers not to do—are our innate ways of processing stress hormones and survival energy and restoring regulation to our nervous system.

The problem is that most of these movements are seen as signs of emotional weakness in the dominant culture defined by toxic grin-and-bear-it masculinity. By suppressing our body's innate responses, we end up trapping hormones and unexpressed survival energy in our bodies, making it difficult to move through the day without triggering a full-body trauma response. It is no wonder that stress-induced illnesses, depression, anxiety, violence, scapegoating, and Othering are rampant and normalized. We lack socially acceptable ways of restoring our nervous system and processing our survival responses.

Relearning to listen to our bodies is incredibly important to our well-being and to the cocreation of new ways of being. It will only become more vital as our social and biological worlds change around us. Disruptions will increasingly affect our lives, and our trauma responses will be further triggered. Now more than ever, we need to cultivate an embodied sense to help in coping with stress, grief, and other uncomfortable emotions.

In these trying times, we must learn to hear our authentic voice—our soma's voice—and allow ourselves to heal and be guided. Our bodies often know how to take care of us and heal, but only if we do not suppress them. So how do we undo our social conditioning and start listening to our wild inner selves? We can start by cultivating silence and contemplation—making room for our inner voice to speak. The next step is not to shush it when we hear it but instead to simply listen to it—even if we do not like what it has to say. The next time an unwanted or uncomfortable thought arises, try not to push it down. Instead, observe. You might try saying to yourself, *Wow. That was a wild thought.* You can then decide to let that thought go. Or maybe you will want to follow it. It just might lead you somewhere interesting.

Mindfulness is a great way to practice carving out silence, but it's not the only way. Moving our bodies is another. Like my dance practice, dedicating time to movement can help us regulate our nervous system and hormones. A regular practice of jogging, yoga, surfing, weightlifting, and martial arts—or intuitive movement more generally—can help clear your body and mind so that you can more easily tune in to your inner self. And, of course, getting outdoors is another way back to ourselves. We can also play a lot more. Children play all the time, until we train them to sit still and resist their impulses. Let's get back to a childlike imagination and play.

Inner work is wild, and it never really ends. It's ongoing and ever changing. It's not an easy journey, but it's a worthwhile one.

EXERCISE: DANCE PRACTICE

Find a space where you can close the door and do not expect to be interrupted for five to fifteen minutes for dance practice. You can put up a Do Not Disturb sign if that is helpful. If you are feeling self-conscious about dancing, you can cover up the mirrors and close the curtains. If you are not self-conscious, maybe invite a friend or family member to join you. Make sure to put on an album or playlist with your favorite songs—the ones you cannot help but dance to.

Try to dance for at least three full songs to help you reset your nervous system and get your stress hormones moving. As you dance, focus on moving the parts of your body where you typically hold tension. Our neck, shoulders, and hips are common areas for storing stress and tensing up. You can start by rotating your neck, shoulders, or hips, but see if you can start listening to how your body wants to move. Does your body want to shake? Stomp? Run in place? Does your body want slow movements like stretching or arm circles? Does your body want to sit down and twist your torso? Or maybe it wants to jump up and down a few times in a row. Maybe for today's practice, shoulder circles followed by neck circles feels like enough, and that is perfectly OK. Regularly check in with your body and see how and for how long it wants to move.

If you find this practice works for you, consider making it a regular routine to keep stress at bay.

Step 5

DEVELOP AWARENESS OF BIASES AND PERCEPTION

> Your assumptions are your windows on the world. Scrub them off every once in a while, or the light will not come in.
>
> —Isaac Asimov

LaUra

Aimee and I like to call this the "Permission to Be Wrong" step. The dominant paradigm tends to glorify the human brain as superior to the rest of the animal kingdom. We think of ourselves as masters of the universe and believe we can discover the ultimate truths about existence. We aim to prove everything through quantifiable means, and anything that cannot be measured is either minimized or excluded as speculation. This is only one way of knowing something. The truth is, our humanness as well as our cultural narratives limit our ways of knowing. Our senses only perceive a limited amount of reality. And the stories we tell ourselves from our upbringing and norms really limit how we perceive and engage with the world around us. Furthermore, our beautifully complex brains are hardwired to make errors and prioritize certain information.

I grew up with an inattentive single mother who would leave my sisters and me alone as children for long stretches of time. Things were not better when she was home—she abused substances and had an affinity for exaggerating, fabricating, or outright lying. As a child whose worldview and livelihood were directly bound up with the messages I received and the reality that was constructed before me, I was often confused. I learned that the only way to detangle the truths from the lies was to find data to support claims. If the claims could not be substantiated, I deduced that they were probably lies. I became a literalist, dissecting language and carefully crafting my messaging. It's no surprise that I pursued biology as my primary major as an undergraduate student. After all, science deals in natural, quantifiable truths. Yearning for verifiable data also helped me become very critical of our culture from a young age. While many were (and still are) stuck in the consensus reality, I began to see through its lies. The dominant culture was not life serving, and my life was an example of this painful reality.

I grew up on the outskirts of a farming community with more cows than people. We did not have a single stoplight, and the village was only about a square mile in size. Because I lacked guidance from my parents, I picked up the sentiments that were expressed openly by the people around me, which, as in most towns and villages in rural Michigan, were pretty conservative.

After graduation, I moved to the city nearby for undergraduate studies, and that fall was the 2004 presidential election. Legalizing gay marriage was one of the ballot proposals for the state of Michigan. At the time, I held a rather pragmatic and literal (read, conservative) understanding that, biologically, being gay did not add up. Sex was a like a puzzle: one piece clicks into another, and that is how babies are made. Forget about love or freedom to choose one's partner. I could not yet embrace queerness or anything other than the norm that I had grown up with. My understanding was that our main goal, as a biological species, must be procreation. In my naive opinion, a marriage between two people of the same sex just didn't make biological sense. So I sauntered into the voting booth and colored in the oval that would keep gay marriage illegal in the state of Michigan. I had a lot to learn.

Sixteen years after filling out that ballot, I had my own wedding. The COVID-19 pandemic was in full swing, and Aimee and I gathered friends and family on Zoom to attend the socially distant ceremony where, on a sunny autumn day, I wed my beautiful wife and fellow heart-centered revolutionary. After the ceremony, Aimee and I performed our first dance together—while wearing rainbow unicorn onesies. All this is to say, it was exceptionally queer.

Every once in a while, I feel a pang of shame for making the decision to vote against gay marriage. When I do, I remind myself that we all have permission to grow and that our thoughts and our understanding of how the world works change as we have more experiences and learn more.

Psychologists have a name for exactly why I did what I did: the bandwagon effect. I hopped on the bandwagon and let the cultural messaging do the thinking for me. There was only one person in my high school class who was openly gay, and my peers did not embrace him well. Everyone else I knew was straight—or so it seemed. As far as I knew, many in my town were against gay marriage not in an openly homophobic way but in the "gay people do not live here," ignorant, homophobic kind of way. Some of my biggest influences practiced Catholicism or evangelical Christianity, proclaiming the sanctity of marriage. I had combined my literal interpretation of biology with the conservative and religious ideas that surrounded me about sexuality and took a seat on the proverbial bandwagon, forming my perspective without really questioning it. It took some time and work to undo those toxic beliefs. A funny thing about being on a bandwagon is you have to realize you are on one before you can hop off. For me, I had to leave my hometown, go to college, and meet Aimee and a lot of other queer folks for me to realize that I had latched onto beliefs that were perpetuating harm.

Groupthink is a common phenomenon among us humans. It's why we are so quick to take to popular beliefs, even when they are irrational, uninformed, or, in my case, discriminatory. The bandwagon effect is considered by psychologists to be a cognitive bias—and understanding that our brain operates with these biases is critical to adequately assessing and coping with the global predicament.

Because we are social beings, we are susceptible to conforming even if it's against our best interest. Think back to Step 1, when we discussed the consensus reality, the shared assumptions and agreements that bind communities in groupthink. Many of our biases keep us tethered to the consensus trance, living in the consensus reality, or the Matrix. Sometimes, we are not even aware of the assumptions we are coming in with. I've learned to examine my assumptions, but I often wonder what other thoughts and beliefs I have been operating with without questioning. What is not even on my radar to consider examining yet? I've shared an example of deconstructing just *one* norm I picked up. No doubt there have been, are, and will be more.

Being alive in a time of the Great Unraveling requires significant humility and compassion. We *all* have permission to learn and grow. Grace comes easier as we embody another of adrienne maree brown's emergent strategy principles: "Never a failure, always a lesson."[1] If we are open, we are always in process. It may take a lifetime to unpack the many cultural norms and messages we receive either directly or subtly. But it is our duty to notice where our limited perceptions keep us from living in alignment with the larger, ecological realities of cooperation, connection, and relationship.

Limitations to Perception

We do not create the space of clear seeing with our usual method.
No questioning, no analysis, no distinctions—just bearing witness to

> what's present. The less we sort, judge, categorize or distinguish, the more we see and feel.
>
> —Margaret Wheatley

We are always only getting a fraction of the story. Our brains are often hard at work, sorting the data that comes in into important and not so important. We are limited by our human senses, and we have a lot of stories about the dominant culture that keep us locked in the consensus trance.

We might not feel it, but right now, our brain is working overtime. Between the sounds and smells of the space we are in, the feelings and sensations in our body as we sit, stand, or touch the pages, the words we are reading and the connections we make involving our memories, cultural upbringing, and personal experiences, we are taking in about eleven million bits of information. And yet the conscious part of our brain can only process about fifty bits of information per second, max.[2]

With so much to take in on limited computing power, our brains often take shortcuts to prioritize which information to take in and process. But sometimes, those shortcuts go wrong. What's more, we are even more likely to make these cognitive errors when we are under stress—and, as we know, we are reckoning with the most complex and pressing predicament we have ever faced as a species. So, yeah, you could say we are under some stress.

The point is, our brains are not always reliable. We need new ways to approach these critical, global problems—we need to learn to analyze our stories about the world and think with our entire bodies. Science shows us that our brains are not the only part of us that can think. Our cells, muscles, nerves, and organs have the power to process information in powerful and mysterious ways. This step is all about how we can think beyond that organ in our heads, tap the insight we hold within our somas, and source from other types of wisdom to face the climate catastrophe.

SENSORY PERCEPTION

We hear, smell, taste, touch, and see but only through our human bodies, which are limited. What we take in through our senses is only a small range compared to what exists in the universe. Animals like cats, bats, and mice can detect ultrasonic, high-frequency sounds that humans cannot.[3] Birds and dogs can see ultraviolet light, and reptiles hunt via infrared radiation. Human beings are limited to seeing only the visible light on the entire electromagnetic spectrum. Many mammals have tactile hairs, or hairs that respond to pressure or touch; humans do not.[4] And there are other animals who can sense Earth's magnetic field or use echolocation; we cannot. These are just a few of the many examples of our limited perception. Of course, we have scientific instruments that can offer insights into phenomena we cannot directly perceive, but these, too, have constraints. We mortals are without omniscience, which provides ample opportunity for inquiring into ourselves, practicing humility, and staying open to change.

THE BRAIN

> Don't believe everything you think. Thoughts are just that—thoughts.
>
> —Allan Lokos

Our human brain is miraculously wired for survival and for making meaning, forming connections, seeing patterns, and countless other things. In every waking moment, this little three-pound organ, about the size of a large potato, is taking in an incredible amount of information—but it cannot process all of it at once. Because of this, our brain makes snap decisions about what to prioritize and will do so automatically and so fast that it will feel like we did not think about it at all. The psychologist and economist Daniel Kahneman calls this *fast thinking*.[5]

Fast thinking occurs when the brain makes automatic choices. Adding one plus two, chewing food, driving a car on a familiar and empty road, pulling our hand away from a hot stove burner, and detecting when someone is hostile are all examples of fast thinking. Our fast-thinking brains take care of both easy tasks and threatening situations when we need to act fast. In that way, fast thinking is the reason humans have survived on Earth thus far—but running on autopilot does have its limits.

These fast responses create room for cognitive biases. Cognitive biases occur when our brain attempts to make a quick, automatic decision. We are all susceptible to them, and often they are the brain's most traveled paths. Generally, they work out for us, and so the brain will use these pathways repeatedly, regardless of whether they are based in reality or not. There's a whole mess of cognitive biases we can fall into—nearly two hundred of them, according to researchers. The table below lists several of the most common. Have you experienced any of them?

COMMON COGNITIVE BIASES

Optimism bias	Overestimating the likelihood to experience positive events and underestimating the likelihood to experience future negative events
Confirmation bias	Looking only at things that confirm your existing beliefs and theories and ignoring things that do not
Hyperbolic discounting	Choosing a smaller, faster reward over a greater, later reward, thus prioritizing the present over the future
In-group bias	Giving preferential treatment to people who are in the same group as you or appear similar to you over people whom you see as different
Status quo bias	Preferring your existing situation or environment and resisting change, even when that change could be beneficial
Bystander effect	Believing someone else will solve a problem, therefore feeling like you do not need to step in
Bandwagon effect	Believing in an idea because others do, without questioning it (groupthink)
Naive realism	Believing you see reality objectively, while others are irrational or uninformed

COMMON COGNITIVE BIASES

Declinism	Romanticizing the past and seeing the present and future in overly negative ways
Pessimism bias	Jumping to negative conclusions, seeing only the worst-case scenarios, and underestimating positive outcomes

Our brain leans on fast thinking whenever it can, because this strategy conserves energy and is often useful. There is nothing inherently wrong with fast thinking. But when it comes to complex issues like enduring the Long Dark, making snap decisions has done and will continue to do more harm than good as it tends to isolate us and focus only on self-preservation. We have to stay in touch with reality, as reality is, not how our biases tell us to perceive it. Part of that means admitting that, in many ways, the climate catastrophe is worse and happening faster than we predicted. Even harder to stomach is understanding that it is a full-blown predicament—a tangled web of power dynamics and economic workings; ecological, psychological, and sociological factors; class and race issues; colonial history; and political beliefs that are all inextricably bound together. We are not arguing for nihilism here. We are saying that, as we face the most immense and pressing predicament of our times, we must learn to intentionally slow down and set aside comforting stories and biases. We have to keep calm, collected, and conscious as we make critical decisions.

So how can we train ourselves to slow down and avoid letting our fast-thinking brains get the better of us? According to Kahneman, the secret to better decision-making lies in *slow thinking*. In our day-to-day lives, we typically use slow thinking to do things we are not familiar with and that require some mental effort. Puzzles, new activities, and complex problems are a few examples of times we use our slow-thinking brains. When we pause to question ourselves or think through a problem, we put the brakes on our fast-thinking brains and sink into a more deliberate state of mind. Numerous studies have proven that the more we slow down to think through a situation, the more rational and disciplined we are in our decision-making.[6]

The most important thing we can do to avoid the pitfalls of fast thinking is to be aware of cognitive biases. Simply knowing the thinking patterns that we tend to fall into when we are under pressure, stressed, or multitasking can help us identify them when they arise. We can also take the time to turn our gaze inward. When faced with a problem, we can ask, *What could I be missing? Is there another way I could be thinking about this situation? Is there anything I do not know for sure about the topic? Am I assuming something that might not be so certain after all?* And the biggest, most difficult question, *Could I be wrong about this?* In pausing to notice and then reflecting, we have the power to change our defaults.

Both slow and fast thinking have their strengths and weaknesses, their places and times. Even when we think more slowly, there are so many other ways that our brains can cause us to make irrational choices. Our habits, emotional states, assumptions, mental illness, cultural stories that live within us, and trauma responses can all lead us to make mistakes and misjudgments. Our brains may be advanced, but they are also limited. In the face of such a thorny and tangled predicament, we have to think beyond our brains. We have to start thinking with our entire bodies.

EXERCISE: GIVE YOURSELF PERMISSION TO BE WRONG

In your journal, write about a time that you changed your mind about a belief. What made you shift your thinking, and how did this realization feel? Were there any cognitive biases at play? Is there a sense of shame or embarrassment about being "wrong" attached to this memory? Have there been times when you judge others harshly for being wrong, making mistakes, or changing their minds?

Be with any uncomfortable feelings that may arise. Then thank yourself for diving into these hard feelings—the work of self-inquiry and growth is difficult. Give yourself permission to change your mind in the future.

The next step in this exercise is to see if you can pause, in real time, when a belief is being acted out. Can you slow down enough to notice and reflect? Is your brain using its default setting? If needed, can you change the course of this particular decision?

THE STORIES WE TELL OURSELVES

The cultures we are born into create a set of collective norms, beliefs, and values that inform our decisions, the questions we ask and those we do not, and our worldviews. We have mentioned the power-over structure within the dominant culture many times throughout these pages, and this way of being causes many of us to believe that some people deserve wealth, safety, and dignity, while others deserve their insecure positions on our planet. These ideas become normalized into the stories that we tell ourselves and each other, masking our perceptions about the way the world works. If we are not willing to explore our cultural values and beliefs and the types of ingrained exceptionalism that live deep within us, we will remain closed off to new ways of being. We must be willing to continuously question our assumptions and the stories we tell ourselves. In *Letters to a Young Contrarian*, Christopher Hitchens argues, "In order to be a 'radical' one must be open to the possibility that one's own core assumptions are misconceived."[7]

With the advent and exponential growth of social media, the corporatization of news outlets, and the outright lies from politicians, it is easy to get confused about the truth. Some say we are living in a "post-truth" world. Many of us have lost our ability to discern truth from opinion, especially when we hear the opinions presented as truths on our social media feeds and by our network news anchors. The rapid production of information and our shortened attention spans have led to a degradation in our ability to think critically and discern important information over the noise of continuous information. In Step 1, we discussed consensus reality and collective gaslighting, both tools that move us further from reality and lead to states of confusion where it becomes easier to follow the masses, people in authority, and our in-group instead of slowing down, discerning, and reconnecting. To minimize feelings of confusion, we generally pick a truth that fits within our cognitive bias, our families, and our culture and use these messages to form our values and beliefs.

THE CALL TO COMPASSION

Because we are human beings with constraints and limitations, we must show ourselves compassion. We have the permission to be wrong and the permission to grow and change as our awareness grows. Our goal in GGN and with this book is to extend that compassion outward and begin to realize that each of us is operating with our own stories, biases, and limitations. Once we can understand that we are all doing our best to survive, we can remove our judgments as others do things we perceive as wrong or misinformed. A compassionate and nonjudgmental nature is essential for connecting throughout the Long Dark.

One step we can take toward compassion is reframing the stories we tell ourselves. Dean LaCoe, a grateful member of the classic twelve-step modality Alcoholics Anonymous, compares the programs: "Like AA, GGN is not a therapy program. It is a story repair workshop. Each of us creates a self-narrative—a 'story' in our mind—that is the window through which we frame all that we see, hear, and feel. Step work is done in small support groups and helps in personally creating and then living . . . a new worldview, one that is bigger, bolder, braver, and more deeply connected to a community. Changing your story helps change your perspective."

EXERCISE: PRACTICE LOVING-KINDNESS

Loving-kindness is a meditation technique in which you focus love and compassion on yourself and extend it to others. Like all meditation techniques, it is a skill that takes time and practice to develop. This practice was inspired by the author and meditation teacher Sharon Salzberg.

Start by carving out a few minutes of silence to bring your attention to your breathing. Do not change your breathing, just bring your awareness to your inhalations and exhalations. Be present in the current moment. Begin cultivating compassion for yourself, your entire soma, and how you are arriving in the moment. You can use a loving mantra to help with this compassion practice. Here are a few examples, though you can use whatever phrases resonate with you.

May I be happy and content.
May I be safe and protected from harm.
Maybe I be enveloped in love.
Maybe I feel well-being and openheartedness.

Repeat these phrases until you feel a sense of ease and loving-kindness toward yourself. This may be difficult the first few times you try it. You may need to practice this several times before you become comfortable with focusing compassion on yourself.

Once you feel comfortable with practicing loving-kindness with yourself, you can begin to expand your compassion to others. In the traditional Buddhist practice of *metta*, or loving-kindness, the compassion extends from yourself, to a loved one, to another being (plant, animal, insect) or element of the outside world, to someone whom you have trouble relating to, and then to all beings everywhere. You can adapt the practice as needed with the following mantras:

> May they be happy and content. / May all beings be happy and content.
> May they be safe and protected from harm. / May all beings be safe and protected from harm.
> Maybe they be enveloped in love. / May all beings be enveloped in love.
> May they find peace and openheartedness. / May all beings find peace and openheartedness.

When we extend compassion toward those whom we have difficultly relating to, we can stop Othering and heal the bitterness within ourselves, making an opening to connect on a heart level.

Body of Knowledge

Within the human body are trillions of cells, hundreds of nerves, and seventy-eight distinct organs, all of which are constantly transmitting information about the world around us and the world within us.[8] In the dominant paradigm, we often think of the brain as the mastermind of the body, the chief commander sending orders to the troops on the ground—our muscles, organs, and cells. But the truth is much more complex. Here is a reminder that 80 percent of the fibers in the vagus nerve serve to bring information from your body up to your brain.[9] The mind-body connection is not just from the top down as many of us think; much of the communication in our body is from the bottom up.

As the Long Dark invites in more uncertainty, we will need the guidance of our whole somas, including our *three embodied knowledge centers*—the head brain, the heart brain, and the gut brain. These knowledge centers provide access to our intuition, our emotions, and our ability to coregulate with others. And it is important to note that all three knowledge centers connect via the vagus nerve, the same nerve that is responsible for engaging our survival responses or the social engagement system we discussed in Step 2.

THE GUT

There is a lot going on in the human gut. Consider this: our gut, which is often called the body's second brain, has as many neurons as the brain of a house cat.[10] Over thirty million different neurotransmitters are produced in the gut, including 95 percent of the feel-good chemical, serotonin.[11] The digestive system is astonishingly complex, and because of the large number of neurons it holds, it can do its work digesting and processing nutrients even when it has been severed from the primary nerve that connects it to the brain.[12]

In addition to the sheer number of neurons and the ability to produce neurotransmitters, the gut is composed of more microbes than human cells. The gut has 150 times more microbiota genes than the human genome, and it is made up of trillions of bacteria, archaea, fungi, and viruses.[13] Many of the microbes are in direct contact with our brain signals via the vagus nerve, receiving and sending information about stress, anxiety, and our state of happiness.[14] Researchers call this the *microbiome-gut-brain axis* and have shown that much of our physical and emotional well-being is dependent on the health of our gut biome and the diversity of microbes that exist. The more diverse our gut biome, the better our well-being. It is a good time to get quiet and listen to our gut for its visceral sensations because it provides key information about our inner worlds.

THE HEART

Rollin McCraty is the research director at the HeartMath Institute, a nonprofit organization that helps people bring their physical, mental, and emotional systems into balanced alignment with their heart's intuitive guidance. In his book *The Science of the Heart*, McCraty asserts that, like the gut, the heart also has a brain of its own that is capable of influencing our perception, emotional experiences, and mental processes.[15] Complete with a complex system of neurotransmitters, ganglia, proteins, and support cells that function similarly to the brain, the heart brain is able to learn, remember, make decisions, feel, and sense—all on its own. The HeartMath Institute has also found that the heart is more than just independent. This organ actually sends information and commands to the brain, and the brain obeys. Our hearts do this via secreting neurotransmitters and hormones like oxytocin, norepinephrine, epinephrine, and dopamine.

And here is something really amazing: our hearts interact with each other. In a study where mothers gazed, cooed, and smiled at their babies while sitting across from them, the mother's and infant's heart beats synced up.[16] It is an incredible phenomenon, and we have no idea how it works. The heart is more than just a pump; it is a place of emotion and of connection. Keeping our hearts open is key in these times.

We call our movement the heart-centered revolution because it is vital that we, as individuals and members of the collective, respond to these disruptive times with openness, connection, and curiosity. The heart excels at all these things. We suggest leading from the heart, the brain situated between the head and gut brains, and integrating the wisdom of each of the knowledge centers.

INTUITION

Intuition is an inner sensing, a type of knowledge or information gained without reasoning. Western researchers do not exactly know where intuition comes from, but studies have found that intuition is not just a myth. In one Stanford study, researchers compared the success of rational thinking versus making decisions based on feelings for a complex decision. The result? Intellectual thinking led to a successful outcome just 26 percent of the time, while affective decision-making yielded a 68 percent success rate.[17] We are more likely to make sound decisions when we tune in to our intuition. But what is intuition, really, and what does it mean to listen to it?

We believe engaging all three knowledge centers with a regulated vagus nerve is key to tapping into our intuition. Our bodies are picking up and processing data in ways that we, in the Western world, have only just begun to study, let alone understand. Our head, heart, and gut brains have intelligence and ways of thinking that inform each brain and our overall soma. If we don't slow down enough to hear the messages from our soma, we might miss important information, ignoring our intuition. Through practices of rewilding ourselves, pausing, calming our vagus nerve, and connecting our three brains, intuition emerges, offering embodied wisdom.

While we in the dominant paradigm are beginning to dethrone the head brain from its importance over the rest of the body, other cultures have understood the interconnected wisdom of the body for thousands of years, as evidenced in the meridian system in Chinese medicine, the three worlds for Andean healers, and the chakra energy centers in Hinduism. Often these ancient traditions add a spiritual aspect to our innate wisdom, which is sometimes called the mind-body-spirit connection. This mysterious source of knowledge may be called embodied wisdom, connection to source, the collective unconscious, or the higher self. Whatever its name, one thing is for sure: our intuitive abilities are powerful. Listening to the innate knowledge of our bodies is a valid way to collect more information about the world around us so that we can make grounded decisions. This benefits our own emotional well-being and helps us to cocreate truly life-centered paradigms as the current ones disintegrate. As the confusion grows around us throughout the Long Dark, our intuition and embodied wisdom will be a continuous source of information on the path.

Letting Go

> I abandon all that I think I am, all that I hope to be, all that I believe I possess. I let go of the past, I withdraw my grasping hand from the future, and in the great silence of this moment, I alertly rest my soul.
>
> —Howard Thurman

Grasping onto our thoughts and beliefs distracts us from noticing openings. It prevents us from being nimble and in the present moment. Learning to let go and flow allows us to develop nonreactive ways of being, meaning that, with practice, we can cultivate a moment between something that triggers us and our response to the triggering.

We cannot control what comes into our lives, but we can begin to control our responses to what comes into our lives. This is the work of the second arrow from the Buddhist parable introduced in Step 3. Through slowing down, becoming more embodied, and feeling our feelings, we can create a moment between an event and our reaction to said event. We are animals who create patterns to survive. Our reactions are not random; we react and repeat successful survival patterns again and again until they become conditioned. Without intentional work to undo these patterns, they are our go-to, unconscious reactions to stressors and look a lot like personality traits. For example, I (LaUra) am quick to anger. If something scares me or makes me terribly sad, my conditioned tendency is to protect myself with the heat and fire of anger. Through years of slowing down, personal healing, and practices that help me feel and process my feelings and trauma, I am able to catch myself before erupting. Not always, but I am practicing. The more we become aware of our inner worlds—our conditioned responses and the places we hold tension—the more we can anticipate and alter how we respond.

EXERCISE: PRACTICE LETTING GO

In this exercise, we will practice bringing our attention to the tension in our body, the stories we might be ruminating on, and our expectations and worldview. We will practice letting go of them. Some of this work may induce feelings of helplessness, fear, or overwhelm more generally. You can always move in and out of the intensity of the practice, backing off if it feels too intense. You can come back to the present moment with your breath and recall that feelings come and go. You are safe in the present moment. Our goal is to build your distress tolerance but not push you past your edges. Only you can know where your edges are. You are welcome to practice this exercise with a trusted loved one or healer for additional support.

To start, clear some space in and around you. Take three long-exhale breaths. Bring your attention to your inner world. Notice where you are holding tension in your body. Are your shoulders raised? Is your brow furrowed? Is your jaw tight? Are you clenching your belly? Portions of your back? What does the tension feel like, and where is it? Breathe compassion into that place in your body, remind yourself you are safe in the present moment, and let go of that tension. Scan your body for the next site of tension. Is there anything else you can let go of?

Notice the stories you are telling yourself. Are you grasping or ruminating on any one story? Did someone say something to or about you that you are holding on to? Is there a judgment that keeps returning? Challenge the story you are telling yourself. Notice if there is a place in your body that constricts when exploring the story or judgment. Practice letting go of the tension in your body and then the situation. Breathe compassion into the present moment. Breathe out the reminder that you can let go of this story or judgment and the anxiety and tension that comes along with grasping onto this situation. Maybe it helps to imagine the story floating away on a cloud or down a river. Are there other anxiety-provoking stories that are living in your body in the present moment? You can notice them and work to let them go, too.

Notice the expectations of the future or the worldview you are grasping onto. Are you holding on to a notion of how the world ought to work? Notice if there is a place in your body that constricts when exploring your expectations and worldviews. Practice letting go of the tension in your body and then the expectation or that aspect of your worldview. Place your attention on the sensation, name the emotion, and come back to the work of letting go. It is OK to experiment with your thoughts and let them go. You are welcome to try on new stories of the future, new worldviews. Let go of your constrictions. This is a practice.

Breathe out, reminding yourself that the future has not come, and we are creating it in this moment. There is courage in letting go of everything we know and feel. This opens us up to more nimbleness, noticing, and the creation of new stories. Spend some time reflecting on what that experience was like for you.

As you move about your day, begin to integrate the practice of letting go—of tension in your body and of thoughts, judgments, expectations, and worldviews.

Finding Clarity

More than ever, we are becoming aware of how our choices affect other people, other life forms, and Earth as a whole—not to mention our own mental health. Even small, day-to-day decisions can feel like major crossroads as we are faced with *impossible choices*. Sometimes, we might get stuck in our attempts to minimize harm to people and our planet in these deeply dysfunctional systems. Which brand of pasta should we use for dinner if we want to be "Earth friendly"—an organic brand that reduces the impact of farming or a conventional brand that has less plastic packaging? Where should we read our news, and how can we tell what's fake from what's real? How can we identify when media is designed to distract, misinform, or manipulate us? Which public figures can we trust to have the collective best interests in mind? What movements, technologies, and policies will pull us toward a more just, life-centered future? How should we invest our precious time and effort to make the most meaningful change or at least steady ourselves for the long haul?

These endless choices can feel overwhelming, but there are ways we can make them more manageable. We will not always make the best decisions—remember, we all have permission to be wrong!—but we can engage a full range of resources to help us on our path. There are numerous techniques to help us slow down so that we can actually hear our intuition and make decisions based on connection and community throughout the Long Dark.

LISTEN TO OUR BODY

Listening to our body sounds great in theory, but how do we do that in real life? What signals should we look for to understand what our bodies are trying to tell us? Everyone's body sends signals in different ways, so we can only speak from personal experience. When faced with a difficult choice, we like to look for two types of feelings: light sensations and heavy sensations.

Light sensations are easy, smooth, and relaxed feelings in the body. These feelings give us a sense of growth and curiosity. Emotions like intrigue, wonder, excitement, or a sense that something just feels right are all good examples of light feelings. Even if we have a little trepidation or fear (what if this decision does not work out?), if our feelings are generally light, it may be a nudge from our body that this decision is the best one to make. Meanwhile, heavy sensations are just the opposite. A sinking feeling in the heart or stomach, sluggishness, dread, tightness in the body, or a feeling that we need to run away can all be signs that something is off about a situation.

There is a caveat to noticing sensations: if we have experienced trauma throughout our life, we might have trouble identifying embodied messages, or the messages sent might be more concerned with protection than with connection. At some point, it was not safe to maintain an embodied connection. Our trauma responses keep our nervous system in overdrive, which interferes with our ability to sense subtle information from the body. As we discussed in Step 2, we can be triggered by events that are reminiscent of a past experience, even if they do not pose a real-time threat. Just like fast and slow thinking, the information sent from our bodies is not always spot-on. If it is difficult to decipher or trust what our sensations are saying, we recommend committing to embodied trauma healing and learning how to reconnect with our body. There are many modalities that can help with this process. Among them, we recommend somatics (e.g., generative somatics, Somatic Experiencing International, or the Strozzi Institute), Internal Family Systems (sometimes called "parts work"), and/or eye movement desensitization and reprocessing (commonly known as EMDR). In the meantime, there are other ways of knowing that we can use to help us make decisions and reduce the ever-present confusion in today's world.

USE SPIRITUALITY AS A GUIDE

LaUra

We are in legitimately daunting and destabilizing times. For a source of wisdom and groundedness and help in surviving current and future upheaval, some folks may look to a spiritual practice. Of course, there are an infinite number of religious and spiritual practices one could engage in. We intentionally avoid prescribing any one, as we believe this is a deeply individualized decision. A spiritual path that works for us may not fit for you. We believe that at its core, spirituality is about connecting with something that is larger than one's individual self. Spiritual practices offer a variety of benefits to our souls, bodies, and relationships and can aid in rooting us in life-centric values.

While spirituality can be used for grounding and increasing connectivity in fragmenting times, there is a danger of engaging in spiritual bypass. The late John Welwood, a trailblazer in the integration of psychological and spiritual work, coined the term *spiritual bypass* as the "tendency to use spiritual ideas and practices to sidestep or avoid facing unresolved emotional issues, psychological wounds, and unfinished developmental tasks."[18] When we use our spirituality as an excuse to gloss over or avoid the very real problems that we are facing, we tend to contribute further to the crisis. Spiritual bypass disconnects us from the larger world and from our own emotions. To avoid increasing disconnection, we can regularly check in about how we are using our spiritual practice. Are we using our spirituality to fortify us with the courage to face our own emotions and the state of the world? Do our spiritual practices help us work toward increasing connection and love? Or, are we using our spiritual practice as an escape to leave this plane of existence and hope for salvation, keeping us from being right here, right now?

Sometimes, being in the present moment evokes fear of our future world. Decades of studying the Buddhist tradition has provided a spiritual framework that supports me when I find myself caught in rumination or fearful projections about the future world. For me, spirituality serves as a reminder of the complexity exhibited through all of life and the larger universe. The fractal patterns observed in the smallest leaf to those in our lungs and the largest river deltas remind me that there is a universal explosion of expression happening around us at all times. The incredible range of colors, forms, sounds, and sizes verify that the goal of the universe is to diversify its expression in patterned ways. I am one part of a complex web of beings all trying to survive. I anchor myself in the embodied awareness that my human manifestation is just one tiny part of a much bigger story that has been playing out for billions of years. Our ancestors are burned-out stars and elements in the water we drink were once in the cells of a dinosaur. When I die, my body will decay, and my atoms will be recycled back into the ecosystems. My deep connectedness helps me in my commitment to minimize suffering throughout my life.

My spirituality is a deep-time relationship with the sacred, a humility that recognizes that my finite, limited experiences and perceptions can never know the whole story of what is playing out. I find equanimity through reflecting and acting on these truths.

EXERCISE: EXPLORING YOUR SPIRITUALITY

Spirituality means many things to many people. Do you have a spiritual practice? What does your spiritual practice look like? Does your practice help you face the predicament with courage? How can it help you discover clarity and groundedness in tumultuous times? What stories of resilience does your spiritual practice teach?

TURN TO COMMUNITY

When our biases and limitations in perception add to our confusion in these disorienting times, it is often helpful to gather community support and guidance. We can turn to trusted mentors, family members, friends, or spiritual leaders and healers for insight and advice. A critical aspect of community is the expanded perspective we often cannot access with our individual perspectives. Organizing individuals into a collective, a community, to come together and help puzzle out a situation or offer guidance is profoundly helpful. Emergence happens in community spaces, with openhearted dialogue. Together, we are able to arrive at ideas we could not imagine while stuck in our own minds. We cannot count the times that turning to members of the GGN or other loved ones has helped us gain clarity and get a more well-rounded perspective. The group consciousness can serve as a much-needed consultant through the Long Dark.

However, we cannot access the clarity of community without some work. Connecting with other people requires vulnerability, and vulnerability is a practice. When we founded GGN, we wanted to create a space that fostered a true community—one where we could create deeper human connections by sharing our vulnerable thoughts and feelings concerning the state of the world. At times, this is discomforting, but we know that these times call us to expand our window of tolerance and endure discomfort. By staying with these uncomfortable, vulnerable moments, we can serve as guides for each other on the path by creating deeper, heart-centered connections. We do not have to know everything; we cannot know everything. A single soma has all sorts of limitations, but three or ten somas working together, talking out the issue and staying open to the range of possibilities, create outcomes that one individual could not imagine. Through a willingness to be vulnerable, trust is created, and relationships are strengthened. This invites new openings and is emergence in action. It also takes the pressure off the individual to know or fix everything.

LOOK TO WISDOM KEEPERS

Lacking adult guidance as a youth, I (LaUra) often looked for perspectives beyond myself to inform the big existential questions. What is the meaning of life? If all life is suffering, what is the point? From the time I had a desktop computer and an internet connection, I spent hours upon hours sifting through quotes and poetry by wisdom keepers, people who seek insights about the human condition and life more generally. Even today, I still create quote books to help me stay grounded, including words from great thinkers, ancient and contemporary.

Though the Great Unraveling is specific to this time in history, the human condition and suffering more generally are timeless and universal. Humans have encountered struggle and enacted resilience strategies for as long as stories have been told. For thousands of years, poetry, art, science, mathematics, philosophy, and religion have been utilized in seeking answers to life's greatest questions. As the suffering of the world intensifies, it is important to lean on wisdom from survivors of conflict and struggle. We can look to ancient texts like the Vedas, the Bhagavad Gita, the Upanishads, the Bible, the Talmud, or the Qur'an or learn from the traditions of Earth's Indigenous wisdom keepers.[19] We can become familiar with our own ancestral history: What are the ancient stories of your people that speak to perseverance and endurance? How did your ancestors overcome their struggles and crises so that you could be here now? What can these stories teach us about surviving through times of darkness?

We can apply ancient wisdom to our modern-day issues and ask ourselves how wisdom keepers or traditions might respond to the predicament at hand. There are many stories available that offer insight for surviving in troubled times. We are seeing a surge of contemporary wisdom keepers, too. It is helpful to think critically and follow a number of teachers to gain a variety of perspectives through the Long Dark. No one person has all of the answers, but there are many ideas, opinions, thoughts, and teachings available to help us along.

EXERCISE: IDENTIFY YOUR PERSONAL WISDOM KEEPERS

Wisdom keepers, historical and present time, have wrestled with how to endure suffering and uncover meaning and joy. When we feel alone or out of ideas and inspiration, it is helpful to look to the wisdom of those voices. There are many guides to help us along this uncertain path. We have some favorite wisdom keepers whom we feel called to learn from. They are teachers who have come into our lives in key moments, whose lessons, books, poems, or essays offer timeless guidance that we needed for our journey.

Using your journal, take a moment to reflect on the wisdom keepers who have helped guide you in your own life. Who are your favorite teachers, ancient or contemporary? What calls you to follow them? How do they see reality? What advice do they have to offer in times of despair? Where do they find motivation to make it through another day?

CHECK YOUR NEWS CONSUMPTION

For many of us, our first instinct upon waking is to pick up our smartphone and make sure society did not burn itself down overnight. When the world feels like it is falling apart, staying up to date on the latest news can feel like a civic duty, a form of activism, a voyeuristic enterprise. But too much news can throw our nervous systems out of whack, sending us lashing out against those we love or reaching for a drink.

Many of us also get our news from our social media feeds, which are not designed to reflect the most trustworthy news and have algorithms designed specifically to our interests and tastes. Sometimes we see good news, but a common phenomenon is to get caught in a cycle of doomscrolling our social media feeds, "looking for the most recent upsetting news about the latest catastrophe."[20]

News, by its very nature, is sensationalized, designed to grab our attention and make us feel something. It is easy to forget that the news industry is designed to make money. Most media in the United States is owned by a handful of corporations, and the whole shebang is funded by advertising (read, more corporations). The more sensational the news, the more people engage, and the more money rolls in. We are not against keeping up with the news, and thankfully there are some ethical and not-for-profit news outlets trying to combat these very issues. But the barrage of news designed to reel us in and rile us up appeals to our fast-thinking brains, stokes our anxiety, and can affect our decision-making.

So how can we tell what is objective and true in a world where information is monetized and targeted toward specific audiences? This is where both slow thinking and intuition come in handy. First, remember that your social media is not a news source. The documentary *The Social Dilemma* exposes how social media feeds are tailored to individuals and designed to show the most outrageous and polarized material available.[21] Whether someone is an impassioned environmentalist or a white supremacist, the content an individual sees on their social media is carefully curated material served up by artificial intelligence, programmed to make each of us to linger longer by playing off our emotions. Our feeds shape how we see the world.

Second, we must question our news sources. If we are only reading news from one website or a single radio or television show, how can we be sure the information relayed is accurate? We can look into who funds the media source. Is there an agenda or a motivation behind how the news is presented? Is the information factual or is someone's opinion being presented as fact?

A final suggestion about news consumption is to regularly check in with our intuition. If something feels off about information we have heard or read, we can investigate further. And if our embodied sense is telling us it is time to turn the news off for the day—listen. The world will keep turning whether we watch, read, click, or disengage. The twenty-four-hour news cycle never takes a break, *so we must* unplug and reconnect with our own reality, notice our breathing, and come back to the present moment.

GROUND IN NATURE

We mentioned the benefits of rewilding ourselves earlier. But there are deeper benefits than helping us relax and reconnect to the natural world.

As our bodies sync up with the natural world, we slow down and begin to take the pace of the more-than-human world. Our frenzied minds naturally calm. From this place, we are able to gain new perspectives, open to imagination, and hear the messages our intuition is trying to send us. In that way, the outdoor world is a necessary space where we can find groundedness in the reality of the present moment.

Try to imagine things from the perspective of beings with minds and senses different from your own. What is a dog's truth? What is the mammoth sunflower's reality? What wisdom does the ant have? What does the mountain, tree, river, or a roaming house cat see? We can try on their reality for an enriched experience. Nature's sense of truth might be much different from our own. Time feels slower, outcomes are not considered, and the present moment is all there is.

EXERCISE: SEEKING DIFFERENT PERSPECTIVES

This exercise can be done as an individual reflection in your journal or as a community discussion. Go outside and spend a few minutes settling into your environment. Take a few long-exhale breaths and start to notice what is around you. What do you see? Hear? Smell? What can you feel around you? Look again and try to identify something new that you have not noticed. Look deeper. Smell something you have not yet engaged with. Are there new sounds you can hear? Take a few moments to reflect on this experience in your journal or with your community.

Next, turn your attention to something near you, maybe a blade of grass or a squirrel. Imagine you are that blade of grass or a squirrel. What might that being feel, see, hear, taste, or smell? What are the needs of that being? What does it long for? Imagine a typical day for that being. What might the grass or squirrel observe that your limited senses cannot take in? Take a few moments to reflect on this in your journal or with your community.

Finally, imagine that you are an ancient being, like a redwood tree or a layer of rock from the Paleozoic period. Imagine what types of change this ancient being might have witnessed. What kind of knowledge does this being have that you might not have access to? How might the ancient one's experience have changed over time? Feel free to be imaginative and playful with this exercise. Reflect on the perspective of an ancient being in your journal or with your community members.

Step 6

PRACTICE GRATITUDE, SEEK BEAUTY, AND CREATE CONNECTIONS

> The more clearly we can focus our attention on the wonders and realities of the universe about us, the less taste we shall have for destruction.
>
> —Rachel Carson

Aimee

It's hard to be a thinking, feeling person in times so rife with loss. I am sometimes bowled over by the amount of suffering already present, and as I envision a pathway through the Long Dark, I often wonder whether I'm strong enough to endure the pain of witnessing ongoing social injustices and species extinction. As someone who lives with depression and anxiety, it's easy to get stuck in guilt or shame or the fear engendered by future projections of a world that has yet to come.

What I've learned about remaining upright is that it isn't always about focusing on the grief, fear, and rage. It's also about balancing these tremendously painful experiences with meaning and joy. Taking in beauty, creating connections, and putting the grit back in gratitude are practices that are gateways into meaning making and joy. And meaning and joy are quintessential in keeping us tethered in an unraveling world.

Putting the Grit Back in Gratitude

Gratitude is all the rage these days. My social media feed is inundated by online articles promising more contentedness through gratitude. Department stores sell glitter-painted and jewel-studded gratitude journals. It is an easy word to throw around and many equate gratitude with happiness. Gratitude is easy, colorful, and fun! Indeed, gratitude might sometimes be those things, but the commodification of gratitude distances us from the harder truth of it. At its core, a gratitude practice is gritty. When we cultivate gratitude in our lives, we are noticing what is here, in the moment. Gratitude thwarts BAU because if we are appreciating what is present, we cannot be overcome by the feelings of scarcity and lack that the advertising industry uses to manipulate us into buying stuff we do not need. When we are grateful, we are focused on what we have—not on what we don't.

Gratitude is considered a practice because in times of suffering, loss, and messiness, it can seem especially trite or forced. It can be a stretch to identify what we are grateful for as the disruption and confusion seem overwhelming. Gratitude does not avoid the pain or cover it up; it serves as a reminder that there is still so much to celebrate amid the chaos and pain.

The potent medicine of grief is a direct reflection of the depths of our love. We grieve deeply because we love deeply, and we can cultivate gratitude for that which we love. It is not an either/or but a both/and. We are capable of holding the complexity of multiple losses and the love and gratitude for the world.

The practice of gratitude is easier said than done, though. I (Aimee) have had a rocky relationship with gratitude that started years ago, when one of my therapists suggested that I start writing down the things I was thankful for. The idea made me cringe. Growing up in a Catholic household, "being grateful" usually came from a place of guilt rather than true gratitude. Guilt was an obligatory price I felt I had to pay for having a privileged lifestyle. I talked about how grateful I felt a lot, and it was because I felt like I didn't deserve the good things I had. I didn't know why I had a safe home and food to eat while so many others didn't. I didn't know why I had chronic health issues and suicidal depression, while my life looked perfect from the outside. Not knowing how to sit with this discomfort, I spent my adolescent years beating myself up for not being grateful enough.

On top of my loaded history with gratitude, it didn't help that when I was depressed, the last thing I wanted to do was write down all the things that ought to make me feel grateful but just didn't. My best friend, my animal companions, the roof over my head—these were things that I thought should stir a sense of gratitude in me. But depression has a way of skewing our logic. As soon as the ink left my pen, my mind rushed to defeat every bullet point on my list: *I'm grateful for LaUra. But then again, we just had an argument. Of course I'm thankful to have a place to sleep at night, but I'm also worried about next month's rent. Especially since we just spent $500 at the vet. Nix the gratitude for our animals.*

Depression already had me feeling like a piece of shit, and each time my gratitude list backfired, I felt like an *ungrateful* piece of shit. Everyone sings the praises of gratitude, so what was wrong with me? Why was I spiraling every time I tried a gratitude list? Was I too far gone? Why was my guilt or frustration always easier to access than my gratitude?

During one of my hospital stays for depression, the therapist asked me to start a gratitude journal. Just the thought of it launched me into a shame storm. But at the therapist's urging, I decided to give it one last try. What did I have to lose at this point?

I sat on the lawn of the hospital with my notebook and started writing the same tired and empty list—LaUra, animals, food, apartment, and so on and so on. Frustrated and out of material, I set down my notebook and looked out at the Wasatch Mountains on the horizon. It was late summer, and the first fall breeze swept the sun's heat and carried it away to the mountains. I felt the cool grass on my fingertips. I picked up my journal and wrote, *I'm grateful for the blades of grass. I'm grateful for the mountains. I'm grateful for this pen. For these symbols on the page that will allow me in a month or ten years to remember what it was like to be here.* I paused. I had spent seven years earning degrees in poetry and creative nonfiction and not once had I stopped to appreciate the absolute miracle of literacy. For the first time, I was practicing embodied gratitude. Instead of writing a hypothetical list of things I felt I *should* be grateful for, I was in the present moment, feeling into my body, discovering the things that brought a bit of joy as I sat on the hospital lawn in Salt Lake City. It took me a long time to realize that a true gratitude practice is not just a lackadaisical listing exercise. It can be a challenging practice, one that often feels impossible in the times we need it the most. But it is a crucial practice because our brains are not capable of holding both gratitude and despair at the same time. Therein lies both the curse and beauty of gratitude. When you are suffering, genuine gratefulness is hard to access. Likewise, when you are immersed in deep gratitude, it is hard to feel shitty.

True gratitude takes grit. It is a practice—specifically, a mindfulness practice. Just like a physical practice, it requires repetition, dedication, and patience. It is an exercise in slowing our thinking, inhabiting our bodies, and noticing our surroundings. More importantly, it is an invitation to feel awe, an excuse to look for something inspiring: the bird in the tree, the vibration of its song carried through the air, the mix of cells, microbes, and energy that I am composed of. To find true gratitude, we may have to look hard, but we will not have to look far.

And here is the best part—gratitude can literally change the way our brain is wired. Gratitude practices are shown to make structural changes in the brains of people with depression in a very specific part of the brain: the medial prefrontal cortex.[1] This part of the brain is not only connected to our emotional regulation and stress relief, but it is also associated with empathy and understanding the perspectives of others.[2] This makes gratitude a revolutionary tool for managing our own uncomfortable feelings in the face of our complex planetary predicament and for connecting with other beings.

After I graduated with my undergraduate degree, I lived in a monastery with retired Sisters of Mercy for a year while I tried to write a book about mental illness and spirituality. One morning, not long after I had moved in, one of the sisters interrupted my writing to say she wanted to show me something. I took my writing time very seriously back then—probably too seriously. Additionally, this particular sister was known for being a bit socially awkward. I could not imagine what she wanted to show me, but I got up and went outside with her. She bent over and asked me to look at the drops of dew on the grass. I did, but I did not know what to say about it.

"Aren't they beautiful?" she asked.

I nodded and smiled. Then she said, "Look at all the colors that the sun makes in just one droplet." And I looked a little closer. I saw that she was right; the drop of water was a microcosm of color that I would have otherwise missed. My appreciation for that moment grows with time. Not only have I stopped more often to admire the dew droplets and remember the sister, but I also remember that she taught me that life is a choice. We can choose to notice the small everyday miracles all around us and be in awe, or we can continue with our schedules and plans, telling ourselves we are too busy to slow down. I choose to notice, to experience awe, as often as I can. To be in awe evokes a natural sense of gratitude and invites us to remember the beauty in the daily routine.

As someone living with depression, I understand the necessity of a gratitude practice, the pausing for a moment to notice what is still here. In my life, I have experienced many losses, and I know that the Long Dark promises ongoing losses—homes lost to storms and sea level rise, species forever wiped away, a stable biosphere altered above our heads.

Now more than ever, it is so important to look toward that which is still present rather than a constant focus on what we have lost or will lose. The grittiness of gratitude invites us to an experience of full aliveness.

JOURNALING EXERCISE: GRIEF AND GRATITUDE

Take three long-exhale breaths and get present in your body.

Draw two columns on a piece of paper. Write the word "Grief" at the top of one column and "Gratitude" at the top of the other.

Spend the next few minutes journaling on all the things you can think of that you are feeling grief about in one column. Spend the next few minutes journaling about all the things you can think of that you are feeling grateful for. The things that you are feeling grief over do not have to be related to your gratitude list. After your lists feel complete, reflect on your answers and compare the two sides. Do you notice any patterns? Is there much overlap between the two lists? Did anything surprise you?

Seeking Beauty

> Finding beauty in a broken world is creating beauty in the world we find.
>
> —Terry Tempest Williams

Like a gratitude practice, seeking out beauty is a countercultural tactic that settles us in the present moment. It is what poets, artists, philosophers, and other wisdom seekers have known forever—and yet, too often, beauty is left out of the conversation about activism. It is deemed a fluffy and soft distraction from the real work of stopping the world from burning. We believe beauty is a tool to help dismantle the dominant paradigm and bring us back into present reality.

Because beauty is subjective and cannot be measured by our reductionist methodologies, it is often seen as superfluous. The creative writing professors in my master's program often told me that the word *beauty* is meaningless and overused. Yet, the photographer and documentary filmmaker Chris Jordan suggests the minimization of beauty by the dominant culture is one of the reasons we are in the current global predicament.[3] In favor of status symbols, pseudo-security, commercialized happiness, and pristine plastic lifestyles, we have given up our connection to the natural world. We no longer see the beauty in a blanket of snow, the miracle of a field covered in trillions of individual snowflakes that are all somehow unique, microscopic miracles never to be repeated. Instead, we see a cold, slushy nuisance that lies between ourselves and our desired destination. Along with our sense of beauty, our empathy and honor for the more-than-human world have been diminished.

Like gratitude, seeking beauty requires us to slow down and observe. It is a mindfulness technique that is always available to us and that we can use for self-soothing—but it is also so much more. When we are enraptured by beauty, our souls open to embrace the larger mystery of the universe. Experiencing beauty entices us to embody feelings of awe and inspiration. The plant that pierces the pavement serves as a reminder that there is a deeper story going on, one that we are a part of, and that the natural world knows and sings to us if we are quiet enough to hear it. Beauty serves as a reminder that life carries on even as we spiral into biological collapse.

A common refrain we hear is that it is often burdensome to appreciate beauty because beauty can remind us of the depth of our sadness over the destruction of species, ecosystems, and populations. A knee-jerk reaction might be to turn away from such strong feelings of awe when they are wrapped up with pain, but this is exactly why it is so important to notice it right now. Every day. As the author, activist, and fierce presence Terry Tempest Williams teaches, beauty is grace, and grace points to wholeness. In a fragmented, disconnected world, we must remember that we are part of the whole. We are connected to everything. Cultivating this sense of wonder and awe deepens our reverence for and relationship with all life. To do this, to live within the paradox of embracing beauty while honoring the pain of destruction, we must employ both/and thinking. We must engage with what is still present and love fiercely even as we face the painful realities of our time.

Good Griever Kasia Stepien says,

> Balancing sadness and beauty is a lasting gift given to me by the 10-Step program. It is a touchstone that I carry with me through the difficult days of this epoch, this time of unraveling, grappling, and reimagining in the Anthropocene. It reminds me to make space for all that I hold dear and am grieving and also for all that I wish to preserve and bring forth into the new paradigms we are all cocreating. I found solace and strength in my own creative and resilient self. I found permission to rest, permission to savor and create beauty, and permission to forgive myself for my own complicity in our society's current trajectory toward climate chaos. I am still grieving and angry. The difference now is that I can hold the grief and anger in one hand, tenderly and with care. And in the other hand, I hold my fierce love for this beautiful world, a comfort in knowing I am not alone, and a faith in the resilience of human beings who just might save ourselves before the end.

Creating Connections

> Rarely, if ever, are any of us healed in isolation. Healing is an act of communion.
>
> —bell hooks, *all about love*

Along with the calming and life-affirming benefits of seeking beauty and practicing gratitude, something remarkable happens as you deepen these experiences of awe: you become more aware of your interconnectedness with all that is. In our capitalist society, we often learn that every person must fend for themselves. Because of this, we often forget just how much we need one another and the more-than-human world.

In ancient Vedic teachings, this sense of connection is described in the analogy of Indra's net. Above the palace of Indra, king of the gods, an unending net of cords is cast, each connected by an infinite number of jewels. Each of these jewels holds the reflection of every other jewel, as well as the entire net of Indra. Even the smallest movement on a single strand ripples throughout all the cords and jewels of Indra's net.

Life on our planet is very much like the net of Indra, for better or for worse. As our planet has warmed, we have become painfully aware that our systems are connected in delicate and unpredictable ways. The negative side of Indra's net is that we get bogged down in how the smallest of our actions will affect other living beings—*I forgot to ask the waiter to hold the straw. Now this piece of plastic will end up in the nose of a sea turtle.* But as we experience beauty and practice gratitude, we begin to appreciate the positive aspects of our interdependence with all that is. We bask in the beauty of a canyon, the softness of the sand underfoot, the millions of years of sediment on which we stand, the cool river water composed of the very same molecules that make up our bodies. Our own bodies have more bacteria, viruses, and fungi living in and on them than they have cells. Our connection with Earth is not theoretical—it is a fundamental truth. We are intimately related to all other beings, including our fellow humans and the universe at large. Each and every atom in our bodies and on Earth was once stardust. Right now, you exist in the particular arrangement of atoms called "you." But these atoms will be redistributed back into the universe after your death and will continue to make new arrangements of all sorts long after.

Embodied interconnectedness strengthens the bonds within communities, families, and all phenomena. It helps us find empathy, compassion, and connection, even with those we disagree with. LaUra and I have found that when our depression screams at us to give up on life, those connections keep us holding on. It is the love we have for our niece-phews, our animal companions, and the dolphins in the ocean that give us the strength to hold on for one more day.

Embodied connection is critical in a dominant culture that silos us with messages of hyperindividualism. From a young age, we are told that if we are successful, it is because we worked harder than everyone else—forget about our cultural conditions, privileges, or lack thereof. In this zoomed-in rat race, only one person gets the job; only one person can be the boss. Meanwhile, there are unspoken stigmas around communal models of living. Shared spaces like apartments, laundromats, day-care centers, and public transportation are often seen, at worst, as a status symbol of the poor and at best as a stepping stone to the coveted single-family home, the private nanny, the Bluetooth-connected washing machine. But the communal resources we shun are often our most sustainable solutions. The paradigms that are to come must involve more communal living with shared tools, machines, vehicles, and of course potluck dinners. How could these small changes make ripples in the infinite and delicate net of our cultures, our climate, and our planet?

We are a collaborative species by nature because we are nature. The dominant paradigm fails to acknowledge this, but as we tune in to our innate sense of beauty and gratitude, our connections with our planet and each other are waiting for us to rediscover them. We are being invited into communion—a homecoming to who we really are. From the air we breathe to the water we drink, we are in dynamic relationship. We are interbeing with all of it. Once we realize this, and protect the world around us as we aim to protect ourselves, we will be more resilient.

We Cannot Forget Joy

> Joy is a radical force because it connects us all to life, and life is enthusiastic for life.
>
> —Rob Hopkins

Aimee

Gratitude, beauty, and a sense of connection all serve to help us remember what makes our lives meaningful and filled with joy. Meaning and joy are a compass, but we need a balance of both to lead us. Prioritizing meaning over joy points the compass toward exhaustion and burnout, and joy at the expense of meaning can result in self-indulgence.

At times, we have been hesitant to ring the joy bell. How can we ask people to embrace joy as we are moving into a time of such loss? It was a lesson we ourselves would have to learn. After reading Anjuli Sherin's *Joyous Resilience: A Path to Individual Healing and Collective Thriving in an Inequitable World*, I learned that "joy is akin to aliveness," and "aliveness comes from having access to and allowing all of our sensations, feelings, needs, and desires."[4] This is a perspective I can get on board with.

Nothing taught me this lesson quite like the COVID-19 lockdown. Despite working full time on the GGN, LaUra and I had not yet figured out a business model to offer us a livable wage. We kept the organization afloat through a variety of side gigs that paid our bills and covered the operating costs. However, lockdown halted our side jobs, and we fell short on our bills.

Eventually, we decided to double down; we pushed harder and harder trying to form GGN into an organization that could provide a livable wage for us. We hosted several additional support groups and special sessions. LaUra and I developed a facilitator training and taught our first twenty students at no cost. On top if it all, we both tested positive for COVID-19, right in the middle of our facilitation training—even after taking precautions to protect against it. Through all of this, LaUra and I worked ourselves beyond exhaustion. We thought the meaning of our work was enough to keep us going. But after a year and a half of pushing ourselves, we collapsed from a lack of joy in our lives. In our pursuit to survive and realize a larger mission, we had not made enough time for celebration, play, pleasure, or fun. And, like so many of us, we had not been able to travel, spend time with our friends and family, or grab a beer at a local brewery—all things that bring us joy or a sense of connection. We learned the hard way that no matter how meaningful our work is, we cannot prioritize meaning over joy.

We fell into the common trap of believing that our meaningful mission of offering something we wholeheartedly believed in was more important than our own pleasure. I wish I could tell you that we now know the right balance between joy and meaning, but every person will have to form their own. Finding that balance is a vital part of this step. What I do know is that in taking time to slow down and notice and then cultivating gratitude, beauty, and connections, we must also make time for joy and pleasure, especially as the losses increase.

Experiences of desperation, hopelessness, and threats to our safety might make it seem that meaning and joy are not available to us, but as long as there is life, they exist. When we encounter such darkness, we must recall concrete instances of beauty, connection, or embodied gratitude from past experiences to help us move through it. Viktor Frankl survived four Nazi concentration camps and focused his life's work on why some people survive great atrocities while others succumb to suffering.[5] In his seminal work, *Man's Search for Meaning*, Frankl returns often to Nietzsche's assertion that those who have "a *why* to live for can bear with almost any *how*."[6] The time is now to identify and nurture our *whys* so that, as the old paradigm fades out and the *hows* seem overwhelming, we are rooted and strong.

EXERCISE: IDENTIFYING YOUR *WHYS*

This exercise is designed to help you center in all that you love and that which brings you joy. You can do this exercise as a solo explorer or share this experience with your community.

Grab a journal. Create some space around you and some stillness within you. Take three long-exhale breaths and then spend a few minutes reflecting on each of the following questions:

- What do you enjoy about your life?
- What is it that you are grateful for?
- What beauty have you witnessed in your life?
- What connections are important to you?
- What and who are the beings, places, items, and things that bring joy or meaning to your life?

Each time you add an entry to the list, take a moment to bring that being, place, item, or experience into your mind's eye. Engage your imagination. What is the thing you just listed? When did you engage with it? What did it make you feel? Is there a texture associated with it? An image? What color is it? Does it make a sound? Is there an image associated with it? Where in your life were you when this experience with it occurred? Bring in as many details as you can about each entry. Pause for a moment of gratitude after each one.

Know you can always come back to this exercise and add to your list.

Take a moment to reflect on the experience. Did anything surprise you about your list? Did any feelings arise while engaging with this practice? If you are doing this exercise in a community, make time for each person to share some things from their list.

Turn to these lists as the despair, fear, grief, or overwhelm more generally threatens to collapse you. These are your *whys* that will help you bear any *hows*.

The Magic of Existence

> Magic doesn't take us out of our ecosystems, it roots us more deeply into them.
>
> —Sophie Strand, "What Is Magic?"

The origin story of our cosmos and life on planet Earth entices us to become captivated by awe, humility, and curiosity. Current research suggests there was a cosmic explosion that occurred 13.7 billion years ago, leading to the formation of our galaxy. Over time, cosmic material condensed to form our planet. The position of Earth just happens to be the perfect distance from the sun for life as we know it to evolve, and the exact angle of Earth's axial tilt allows for seasonality. Billions of years of evolution were suffused with losses, expansions, losses, and more expansions. Life emerged. Somehow, the carbon, oxygen, water, and nitrogen cycles evolved allowing for the distribution of key elements, energy, and molecules. The continents shifted, oceans formed and disappeared, and ecosystems drastically changed.

The *Cambridge Dictionary* defines *magic* as "the use of special powers to make things happen that would usually be impossible."[7] There are elements of magic in life's origins on Earth. Despite all the research and tests, we still cannot reproduce the origins of life in a laboratory. The impossible seems to have happened over billions of years to get us to this moment in time. Maybe coincidence after coincidence occurred, or maybe we are being guided, called, or nudged forward by an evolutionary force.

In a time of planetary destruction, direction, guidance, and meaning might be found in pausing to notice when the evolutionary force, the cosmos, or our ancestors want to guide us. By following our intuition, pursuing meaning, cultivating joy, and noticing synchronicity, we make magic; we allow space for the impossible to emerge.

Carl Jung described synchronicity as a "meaningful coincidence of two or more events where something other than the probability of chance is involved."[8] From our experience, the more we pay attention to synchronicities in our life, the more they occur. While Western psychology can explain this phenomenon through what is known as the frequency illusion, or the Baader-Meinhof phenomenon, we find this explanation incomplete and will remain open to the mysteries of the universe. We see synchronicities as encouragement to pay closer attention to things that reappear in our lives. For example, in a conversation at a party, Brooke Williams, wilderness protector, writer, and husband to Terry Tempest Williams, mentioned to me that Jung was one of the most brilliant minds of the last century. "I haven't thought much about Jung since I was an undergrad," I said. "I'll have to revisit him." A few days later, I was strolling through the Salt Lake City public library and noticed a Jung quote above a door: "The meeting of two personalities is like the contact of two chemical substances: if there is any reaction, both are transformed." It was a timely happening, so I snapped a photo of the quote. The next day, LaUra and I met for tea with our spiritual mentor, Kinde Nebeker, who mentioned she does public art around the community. We inquired to know more, asking what type of public art and where. She said that she picked all the quotes and chose where to place them around the library.

"So you are responsible for the Jung quote outside of the restrooms?" I asked.

"I sure am," Kinde replied.

LaUra and I looked at each other and laughed. Coincidence? Maybe. But when something comes into my view multiple times from a few different angles, I make sure to pay attention. For me, this series of coincidences was an opening. It was all the nudge I needed to start exploring Jung's work again. I went home and dug into *The Red Book*.

We can practice magic by developing a relationship with the evolutionary force, whether we label it as God, the sacred, the divine, or our ancestors. Perhaps magic is just another way of knowing that is often blocked by our biases and perception. When it comes to impossible things falling into place, we can remain open. Much of what we need to do to heal the world and bring people together feels impossible. Just because something feels impossible does not mean it is impossible. There is so much we do not know about the cosmos, Earth systems, and life. There is magic in the everyday just waiting to be witnessed and exalted. Setting the conditions for desired outcomes to emerge is like casting a spell: we set our intentions, create the conditions, and wait for the unknown force that drives life to take over. We can rest in the mystery, believe in our personal and collective power, and look for synchronicities to guide our way.

Step 7

TAKE BREAKS AND REST

> Caring for myself is not a self-indulgence, it is self-preservation, and that is an act of political warfare.
>
> —Audre Lorde, "A Burst of Light"

Aimee

When we first started hosting GGN support groups, we called this step Take Breaks and Rest (as Needed). It was not long before we dropped "as needed," because we realized people (including us!) don't know when they need rest. In the United States especially, we have been culturally conditioned to work hard and, when that doesn't work out, work even harder. We learn to ignore the signals sent by our somas to slow down and take energy shots instead of resting. Because the dominant culture doesn't create space for taking breaks and supporting oneself, especially if we can't pay our bills, we have to find ways to demand breaks and rest for ourselves.

We often say that we created the program we needed not because we practice all these steps to perfection but because we are engaged in the process alongside our support group participants. We acknowledge and create space for our own messiness, confusion, and shortcomings, just as we encourage our participants to do. In our groups, we are relearning how to be with one another without fixing each other. We make room for tremendous grief, despair, fear, rage, or any other feeling to be expressed as each of us shares a portion of our journey.

Sitting with our own emotions is hard work and so is facilitating spaces for people to come with their heavy and painful feelings. For four years, GGN was kept afloat by our idealism and dedication and support from a small number of devoted community members. Yet while I have regular practices that help keep me grounded and embodied, I eventually burned out on facilitating GGN support groups. My entire soma stomped on the brakes by radiating extreme pain throughout my entire body. I had to come to terms with my emotional and spiritual limitations. My basic needs had not been met, and my soma knew it.

The summer of 2021, I was in bed with excruciating pain levels nearly every day. Thankfully, I had received Medicaid a few months prior, but another obstacle quickly became apparent: the little town I lived in lacked sufficient health care services. I was bounced around from professional to professional, gaslit about my suffering, and unable to see a pathway for healing. At the time of writing this, almost half a year later, I am on a path to recovery from chronic nerve pain after numerous referrals, two surgeries, and a couple inpatient hospitalizations.

This story is not finished. I am still healing and creating a balance with the work that is so desperately needed in these times. But I know from past burnouts and mental health breakdowns that it's a lot easier to heal when we haven't collapsed ourselves entirely. The problem is, we often don't see a breakdown coming in time to avoid it. And even if we see it coming, many of us feel the tension that stopping work means stopping paychecks, and we need money to exist in this capitalistic society.

This step, Taking Breaks and Resting, is all about undoing our capitalist indoctrination and practicing self-care—and no, we don't mean spa treatments, shopping sprees, craft cocktails, or other corporate #selfcare schemes. We're talking about break taking as an act of rebellion against these soul-crushing systems and rest as a human right. It's learning how to identify places where we can pause, unplug, withdraw, and find the rest that most of us so badly need.

Self-care may have been co-opted by corporations, but they sure as hell didn't invent it. The practice of self-care is a radical one that has been in the activist's toolbox since at least the sixties, when it was popularized by the women of the Black Panthers.[1] As we continue to face the interconnected crisis brought on by the climate emergency, systemic racism, failing social systems, and more, it is critical that we look back on the wisdom of these activists who knew that rest is not only a requirement for nourishing ourselves but also a form of dissent against BAU.

The Right to Rest

> Every person deserves a day away in which no problems are confronted, no solutions searched for. On that day we need to withdraw from the cares which will not withdraw from us.
>
> —Maya Angelou

You might be surprised to hear that this is one of the most controversial steps in our program, and there is a good reason for that: taking a break has been sold to us as a privilege for those who have worked hard enough to earn it.

Before we dive into modes of rest and break taking, we want to acknowledge that rest is a human right—and like so many of our rights, it is often not afforded to those who need it most. For those working multiple jobs, single-handedly raising a family, caring for elderly parents, making less than a living wage, or for whom taking a day off means losing a day's pay, resting is not easy. Taking a break is often even harder for women, who tend to be saddled with the responsibilities related to family, home, and child-rearing. Moreover, women, trans folx, and people of color are less likely to have access to paid sick leave, adequate maternity leave, paid vacation time, or affordable childcare. Meanwhile, those with disabilities (visible or invisible), who may need more time for break taking and slower, more methodical work, are often discriminated against by employers. The disparity of access to rest in our world is an injustice perpetuated by a system that does not value life—especially the lives of people who are already marginalized in so many other ways.

Any viable path toward a truly just and life-centered world must ensure everyone has access to healing rest. But we acknowledge that right now, this is not the reality. For many, carving out time for work, rest, *and* activism may well be impossible—and this is quite convenient for the power-over structure. When we are weary, stressed out, or too busy with work, we burn out from trying to do too much. We do not have the time or energy to make radical movements happen. We become easy to manipulate. We internalize capitalistic values and project them onto our families and our communities.

In so many activist circles, resting is taboo, working hard is a badge of honor, and exhaustion is etched into one's identity. These are all traits our capitalistic culture values—the very same culture that created the predicament we are facing. We may feel that the overlapping crises are a ticking time bomb, that there is no time for rest. But by creating space for rest when and where we are able, we resist the dominant culture. We may not be able to defuse the bomb, but slowly, calmly, steadfastly, we can undo the culture that built it.

We are not here to tell you to "just take a break" if you cannot. But we are here to call bullshit on this rigged and inhumane game. As the systems that deprive us of rest are reimagined, dismantled, and rebuilt, we must advocate for ourselves and our communities. With this step, we hope to provide new modes of break taking and self-care and to envision a future where we can all afford to slow down and take a nap—or two or three.

The Radical History of Self-Care

The term *self-care* was coined in the 1950s by medical experts who were attempting to build autonomy and positive habits for institutionalized patients.[2] It was not until the late sixties and seventies that the idea of self-care began to gain traction in the larger community, all thanks to the women of the Black Panther Party.

Along with fighting for the liberation of marginalized groups, the Black Panthers were truly incredible community organizers. Much like today, the systemic racism of the time meant Black people had little access to adequate health care, childcare, education, and social services. The Black Panthers took matters into their own hands, organizing community youth programs, free health clinics, breakfast programs, food banks, grassroots grade schools like the Oakland Community School, and even a People's Free Ambulance Service.[3]

The Black Panthers were pioneers of community care, but the organization also understood the importance of individual care. Perhaps the most outspoken advocates of self-care in the Black Panther Party were Angela Davis and Ericka Huggins, both of whom devoutly practiced yoga and meditation while in jail. Davis recalls in a 2018 interview with Afropunk that most activists at the time were not thinking about caring for themselves corporally, nutritionally, spiritually, or mentally[4]—and sadly, we see this all too often in activist circles still today. Both Davis and Huggins recognized that self-care was not just an individual practice—it had collective value. Not only does it allow us to bring our entire selves to the predicament, it also protects against activist burnout. Seeing the value of these self-care practices in collective spaces, Huggins began teaching yoga and meditation to other party members, including Black Panther Party leaders Huey Newton and Bobby Seale, as well as prisoners across California. In 1968, FBI director J. Edgar Hoover called the Black Panthers the "greatest threat to the internal security of the country."[5]

We think he was right. The Black Panthers created their own modes of care, both communal and individual. In doing so, they refused to participate in BAU, and that is the ultimate threat—to withdraw from the system, to refuse to uphold it, to care for ourselves and each other when our systems fail to do so.

By the eighties, the Black Panther Party had dissolved. In her interview with Afropunk, Angela Davis questions whether the movement might have lasted longer had they all been practicing more self-care. But even as the party disintegrated, the idea of self-care gained momentum in civil rights and feminist groups and continued to be echoed by Black feminists throughout the seventies and eighties. Self-care evolved from a medical movement to a political one designed to fortify the soul against a racist patriarchy. One of the feminists who most famously carried the torch of self-care advocacy was the writer and civil rights activist Audre Lorde. We began this chapter with a quote from Lorde that we turn to often for inspiration: "Caring for myself is not a self-indulgence, it is self-preservation, and that is an act of political warfare." Lorde wrote these words in 1988, after discovering for the second time that she had breast cancer. As a Black, lesbian woman, the refusal to participate in the systems rigged against her was truly radical. Lorde's writings are a poignant exploration of rest as resistance, of the systemic injustices that deprive break taking from those who need it most, and a reminder that often, doing something for oneself is the most powerful thing we can do for the collective.[6]

Self-care has since been co-opted by corporations, advertisers, and influencers, particularly in the age of social media. On Instagram, grinding and side hustles are glorified with images of pour-over coffee and bougie tech offices complete with beer bars and ball pits. Meanwhile, corporations bath bomb us to death with ads for the perfect #selfcaresunday products. (Incidentally, these luxurious ads are often made by employees working fifteen hours a day in hopes of climbing the corporate ladder.) But capitalism has far from crushed the spirit of self-care that

Black feminists brought into being. One of our favorite thought leaders in this space is Tricia Hersey, who founded the Nap Ministry. Based in Atlanta, Georgia, the Nap Ministry explores the power of naps as resistance through performance art, workshops, and creating sacred spaces of rest within the Atlanta community. In Hersey's words, "We don't want a seat at the table. F#$k the table. The table is full of oppressors. We want a blanket and pillow down by the ocean. We want to rest."[7]

In our current reality, it is not realistic to completely withdraw from the dominant paradigm all the time. Unplugging from the system, for so many of us, means not feeding our kids or having housing. But by practicing self-care, we can build up resistance bit by bit. We link up with community members and create strong networks of care and aid. We look for openings to take breaks and withdraw from the unrealistic expectations placed on us by the toxic dominant culture. And by imagining new ways to work and rest, we can create a future where we depend less on broken systems for our well-being and more on ourselves and our communities. The beauty of the intersectional theory of self-care that Audre Lorde, Angela Davis, Ericka Huggins, Tricia Hersey, and so many other Black activists have brought into our remembering is that when marginalized groups are liberated from oppression, we will all be liberated. Those of us in the consensus trance have forgotten that liberation is a holistic endeavor. None of us are free until we are all free. Self-care and community care are dependent on each of us having agency over how we use our time, bodies, and energy.

More Being and Less Doing

Frenzy, fear, hopelessness, and trauma responses keep us from engaging our imagination. These overwhelming feelings and wounds keep us living small, constricted lives. There is no time for expansion, connection, or imagination if we are convinced our very survival is at risk. And, expansion, connection and imagination are exactly what is needed now.

We speak often about openings, about a third way when the tension is too much and we cannot exit (see page 77). Openings are noticed through imagination. We need space and clarity. We need idleness and equanimity. We need creativity.

Doing and *being* exist on a spectrum. The dominant paradigm glorifies *doing*, often at the expense of our health and needs. Messaging from the power-over system has convinced us that to survive, we must constantly *do*. Our livelihoods are bound up with our production. If we cannot produce, we will die. *Being* is often seen as a distraction, navel-gazing, a waste of time, and an interruption to our levels of productivity. If we prioritize *doing* over *being*, we become cogs in the power-over machine, which could potentially lead to burnout. But if we focus on *being* over *doing*, necessary tasks will not get accomplished.

Each of us can practice being *and* doing in a balanced way. LaUra is more of a *doer* than a *be-er*, and Aimee is more of a *be-er* than a *doer*. We all have a unique blend and comfort level on our doing and being spectrum. This is due to our trauma responses, personalities, life experiences, and, as mentioned a moment ago, our culture. What combination do you notice about yourself? If you are more of a be-er, how can you balance your need to just be with doing the tasks necessary for what you wish to accomplish? If you are more of a doer, how can you create more space and time around you for more being, more intentional slowness, and more proactive rest?

Regardless of where you land on the *being-doing* spectrum, slowing down allows all of us to hear the aches, pains, knowledge, and desires of our bodies so that we can nourish ourselves and carry on *doing* the work we love. It gives us time to unfurl our tightly wound brains and to sink into the type of slow thinking we need to endure the predicament we are in. Consciously slowing down, taking breaks, and resting allows for clarity of mind, a calm nervous system, and holding the tension of the time. Taking breaks gives us time to play, and play opens us to imagination, spontaneity, passion, and the birth of new ideas and ways of being. When we connect with a well-rested community, novel ideas and ways of being have a chance of being implemented.

As we cultivate the heart-centered revolution, applying the capitalist values of grind culture to our work will not get us there. Instead, we must go slowly, calmly, and rested. Moment by moment, we can carve out openings for new paths to emerge.

Expanding the Idea of Rest

> Rest is not idleness, and to lie sometimes on the grass on a summer day listening to the murmur of water, or watching the clouds float across the sky, is hardly a waste of time.
>
> —Sir John Lubbock

What can we do right now to nourish our souls and restore our energy? We can start by questioning our questions. That is, we should ask ourselves not if we are getting enough rest but if we are getting the right *type* of rest.

Many of us have taken a cozy midday nap only to wake up feeling more exhausted. This phenomenon is so common that Dr. Saundra Dalton-Smith, board-certified physician and author of *Sacred Rest: Recover Your Life, Renew Your Energy, Restore Your Sanity*, made it her mission to uncover why taking a nap simply is not enough to make us feel restored.[8] What she found is—get ready for this life-changing fact—sleep is only one of several types of rest. According to Dalton-Smith, there are seven distinct areas of our life in which we need rest: physical, mental, sensory, social, emotional, creative, and spiritual. We can sleep all we want, but if we are not intentionally identifying and tending to the types of rest we need, then all those naps are for naught.

Now, don't get us wrong—getting a good night's sleep is imperative to our well-being. Sleep deprivation has been linked to a heap of health issues from mood disorders and increased stress to high blood pressure and even cancer.[9] But the problem with focusing on sleep as the only type of rest is that we tend to think it is the only solution to our exhaustion, when, in fact, it only attends to the first type of rest we need: physical rest. Physical rest needs to be achieved in two different ways: through passive types like sleep and napping and through active ways like mindful stretching, yin yoga, or a slow walk.[10]

The second type of rest we need is mental rest, and it is especially important in our nine-to-five, forty-hour-work-week world. Between doing our jobs, going to school, caring for a family, keeping up with the news, and change making, we spend immense amounts of mental energy every day. A lack of mental rest might manifest as irritability, tiredness, an inability to focus, and lack of motivation, and aside from physical rest, it is probably the type of rest we are all the most familiar with. We can restore our minds by taking breaks throughout our day and intentionally limiting our screen use and news consumption. Of course, meditation and mindfulness fit into this category too! We can also do an activity that is purely for fun, such as taking a walk through a forest or even watching a favorite TV show. But be careful with that remote—vegging out is not always the type of rest that is needed, and it could even lead to a lack of sensory rest.

Sensory rest is essential in today's hyperconnected, information-overloaded paradigm. When we constantly bombard our senses with social media feeds, text messages, podcasts, TV shows, and audiobooks—however informative and well meaning—we do not give our bodies and minds enough time to unwind and rest, and we do not make time for intimate social connections. Dalton-Smith says not only can a lack of sensory rest show up in neck tension and tired eyes, it can also manifest as broken relationships.[11] The ultimate sensory rest would be spending time in a sensory-deprivation tank, but that might be expensive and difficult to access. Another way to rest your senses is to unplug from electronics in your free time. Putting our devices in "phone jail" before spending face-to-face time with loved ones is an important way to minimize distractions. We can also create intentional, mindful, sensual experiences, like smelling flowers, drawing a lavender bath, sipping herbal tea, giving or receiving a massage, or incredibly present sex.

Identifying "life-giving people" is how we achieve social rest, according to Dalton-Smith.[12] Who are those folks that do not need anything from you and in whose company you can just be you? The answer varies from person to person and depends where we are on the introversion/extraversion scale. Maybe social rest looks like quality alone time for you or being surrounded by a small group of like-minded people. The goal of this type of rest is to find spaces to be your whole self without having to give anything.

If social rest is a struggle, Dalton-Smith suggests we might be suffering from a lack of emotional rest. We all need space in which we can feel our emotions authentically and unapologetically—which is exactly why GGN support groups exist. When we stuff and avoid our emotions to please or appease other people, we are not getting the emotional release and rest we need. Spending time with supportive loved ones, a trusted healer, or a community space where we feel comfortable showing up authentically is the key to restoring our emotional well-being.

Another type of rest we need is creative rest, and it is not just for artists and writers. We use our creativity when we plan a birthday party, write an important email, or problem solve at work. Creative rest is not about creative production—painting or writing a poem may not make us feel more creatively rested but less. Instead, replenishing our creativity is a passive act of appreciation and awe. It can also be "allowing white space in your life and giving room for your creativity to show up."[13] Additionally, creative rest could be reading a novel for fun, filling our workspaces with artwork we enjoy looking at, visiting a gallery, or going on a hike to soak in the beauty of the natural world.

If life lacks meaning and it feels like we are just going through the motions, we may be in need of the final type of rest: spiritual rest. Nurturing our spirit does not necessarily mean meditating or going to a religious or spiritual service, although it can mean that. Spiritual rest is about taking the time to discover purpose, meaning, and belonging in our lives. This can mean so many different activities for every different form of spirituality, such as going on a hike, playing with animal companions, doing meaningful work, talking to a tree, or spending time in supportive community spaces.

EXERCISE: GET THE RIGHT REST

Next time you find yourself feeling burned out, exhausted, or lacking motivation, wait one moment before you pick up the remote, scroll your phone, or hit the sack, and instead take some time to identify and tend to the type of rest you really need. What are you missing in your life? Practice a new way to replenish yourself that is not your habitual go-to and reflect on what that experience was like for you.

TYPES OF REST	WAYS TO REPLENISH[14]
Physical	Take a nap. Eat a balanced, nutrient-rich diet. Drink plenty of water. Practice good sleep hygiene—no screens in bed, regular sleep schedule, eight hours of sleep. Practice mindful stretching, yin yoga, slow walking.
Mental	Take regular breaks from social media and electronics. Take regular breaks from the news. Create a space for silence. Schedule short breaks throughout your day. Go on a mindful walk. Practice meditation. Do something just for fun: color with markers, watch a stand-up comedy special, cook a meal, and so on.

TYPES OF REST	WAYS TO REPLENISH
Sensory	Turn off electronics, including background noise like TV, the car radio, or podcasts. Spend time in a sensory-deprivation tank. Practice sensuality—draw a warm bath, dig your toes in the sand, feel a cool breeze, engage in mindful sex, or masturbate. Give or receive a massage. Use essential oils. Mindfully eat your favorite food.
Social	Balance interactions on social media with face-to-face time. Make connections with like-minded people who share your passions and interests. Spend quality time with loved ones. Call an old friend. Identify and connect with "life-giving" people.
Emotional	Practice emotional honesty and integrity. Foster connections with people who do not expect you to perform and are open to listening to your honest feelings. Establish a regular journaling practice. Make time to feel all feelings and allow for embodied emotional releases (cry, shake, shout, or jump).
Creative	Color, draw, or draft a poem. Read a book for fun. Go to the theater or listen to music. Appreciate beauty. Spend time with the more-than-human world. Visit new or unusual places. Rearrange your furniture. Create space for pauses. Try a new creative endeavor—without pressuring yourself to be "good" at it.
Spiritual	Meditate/practice mindfulness. Pray. Read, watch, or listen to spiritual wisdom keepers. Spend time in awe of the more-than-human world—a flower, a river, the view from a window. Connect with a kindred spirit. Attend a meditation/mindfulness group or religious/spiritual service.

Be a Part of the Band

Creating and maintaining meaningful connections with others is a vital part of our individual well-being. Having a community is important—even if it is a small one.

A strong community is a lot like a band. A variety of individual instruments come together to create a song. But not every instrument is playing all the time. Part of what makes music such a unique sensory experience is that some instruments rest while others carry the melody. After a rest, an instrument plays a series of notes and adds a unique zest to the song. The rest allows us to enter back into the song and put our full and vibrant selves to work.

Even if we are part of the band, resting can still be a challenge. As we have mentioned, taking a break can be difficult for those of us called to activism or other types of revolutionary, healing work. Resting is further complicated because our power-over structure does not allow us to easily create communities that can care for us and hold us in all types of circumstances. Conventional norms keep us very individualized in our struggles. Many of us do not know how to ask for help or really engage in community care. Similarly, we are taught to privatize money, food, and care, allocating them only to our immediate family. How do we move from feelings of scarcity and competition to cooperation and sharing so that we might begin trusting others to carry the melody if we need to take a step back? There are models that we can look to for community care and mutual aid, some of which we mentioned in the community section.

The work of the change maker or healer will never be done—especially during the Long Dark. Our dominant paradigms need dismantling, diversifying, and restructuring as soon as possible to protect our living systems on Earth. This is when we go to work dreaming, visioning, connecting, and, of course, undoing and unlearning. Many of our fellow change makers and healers feel this sense of urgency in their bones. Radical transformation of our systems and worldviews takes time and energy. We cannot continually show up without creating space for joy, celebration, and pauses. Burnout is a real threat to change makers and healers everywhere. But when we connect with the band, we remember that we are part of a larger melody. There is time to play our instrument, and there is time to rest. The work will be waiting after the break.

It Is Time We Talked about Meditation

> The practice of meditation is not a passive, navel-gazing luxury for people looking to escape the rigors of our complex world. Mindfulness and meditation are about deeply changing ourselves so that we can be the change that we see needed for the world.
>
> —Larry Yang

Practicing meditation is one of the most radical self-care tools available to us. While it can help us feel deep relaxation, it also has the capacity to improve our well-being and cultivate equanimity in tumultuous times. This practice also teaches us how to be less reactive, even in difficult situations. Like all practices, meditation requires time, intention, discipline, and energy.

Meditation is one of the single most effective grounding practices we can use to calm our minds, feel our feelings, stop identifying as our thoughts, and, most importantly, deal with the chaos that lies before us. As the world continues unraveling in unpredictable ways, our nervous systems will be naturally inclined to respond with our survival responses. We know from previous steps that when our survival responses are in full force, we do not have access to the best versions of ourselves. One of the many benefits of meditation is that it moves us from conditioned responses by literally changing the shape and function of our brains.

The table below provides the results from four studies that have shown how meditation changes the structure and function of the brain.

STUDY	BRAIN REGION ANALYZED	FUNCTION OF THE REGION	CHANGE OBSERVED	POTENTIAL IMPLICATION
"Meditation Experience Is Associated with Increased Cortical Thickness"[15]	Prefrontal cortex	"Control panel of the brain"; responsible for high-order cognitive processes like reasoning, decision-making, social cognition, and personality expression[16]	Increase in the thickness for longtime insight meditators	Increased capacity for compassion, social connection, and rational thinking
"Mindfulness Practice Leads to Increases in Regional Brain Gray Matter Density"[17]	Hippocampus	Involved in memory processes, behavioral regulation, and spatial cognition	Increase in the gray matter after eight weeks of mindfulness-based stress reduction	Increase emotional regulation, memory, and spatial relations
"Impact of Short- and Long-Term Mindfulness Meditation Training on Amygdala Reactivity to Emotional Stimuli"[18]	Amygdala	Threat detector; assigns emotions like fear or anger to outside stimuli; triggers survival responses	Training in awareness-based compassion meditation led to reduced anxiety and activity in the amygdala while experiencing uncomfortable emotions; decreased activity carried over into the nonmeditative state	Reduction in stress, fear, and anxiety; the ability to tolerate distress instead of instantaneous reactivity
"Meditation Experience Is Associated with Differences in Default Mode Network Activity and Connectivity"[19]	Default-mode network (medial prefrontal and posterior cingulate cortices)	Our wandering mind; responsible for introspective thoughts and creating our self-referential thoughts	Decrease in activity	Reduction in ruminating, obsessive, or worrying thoughts; ability to catch when one's mind wanders and bring it back to the present moment; less focus on "me"

Numerous other studies have also exalted the many benefits of a meditation practice, but the main point is that as we regularly practice meditation, we become more compassionate, more patient, and less reactive in our daily lives. We are also less stressed and have lower blood pressure, slower heart rates, and less pain. Occurrences of depression and anxiety are reduced, and our quality of sleep, attention, and memory tend to increase.[20] A sustained meditation practice makes nuanced responses to chaotic situations more readily available to us. Meditation can also open us to responses that are less individualized and more cooperative.

Remember: what we feed grows. This is true for our neural pathways as well. Neural pathways are the connectors between one area of our nervous system and another, and our thoughts and feelings travel along them. Just like worn paths on a forest floor, the neural pathways most traveled are the easiest to take—they are well-beaten and clear of debris. New or unusual thought pathways, however, don't yet have a path in the dense forest—they are covered in brush and loose stones and are more difficult to travel. When we are under stress and our brain needs to make a quick decision, it will take the most traveled neural pathway (our conditioned response), which is not always the best one. Our brains get stuck in routines, which cause us to react the same way time and time again. Through repetition and by practicing meditation regularly, we create new neural pathways and clear the brush and stones on the newly created path, offering ourselves new opportunities to respond. We go inward to help us change the outer world!

Find What Works for You

LaUra

For some, *meditation* is a loaded word. Much like a gratitude practice, a meditation practice is something that many people feel they should do but do not do often enough or well enough or the "right" way. There is a common misconception that there is only one type of meditation: the one that requires you to sit upright and still, clear your mind, and focus on your breath. However, there are many ways to meditate from a variety of schools of thought and religious and secular practices. We find that seeking beauty and embodying gratitude are their own form of mindfulness meditation. They invite us to be present with what is, to slow down and notice the current moment without judgment. They are practices unto themselves, and they can be a friendlier entry into meditation than some other forms.

As a trauma survivor, I have had a difficult time with stereotypical "sit in silence and come back to your breath" meditation. It usually leaves me feeling overwhelmed or on edge rather than centered and calm. I've tried to push through and practice more and harder and longer, and it simply doesn't work. Here's why: When the body is too painful to inhabit, such as during a traumatic event, we "leave" and sever our connections to our sensations. It's a critical survival technique as we're experiencing a trauma, but if we don't process our overwhelming feelings and built-up stress hormones after the event has passed, the feelings and energy get stuck; our brain makes up its mind that the body is a painful place to be and disconnects from the bodily sensations. And if this happens over and over and over again, like in my case, it becomes difficult to reintegrate the bodily sensations. The result: Reconnecting with my body through meditation often ends up in extreme discomfort from the tremendously painful feelings and trauma stored deep within my soma. The surge of unfamiliar and uncomfortable sensations and feelings is too much, and my mind checks out of the practice.

The author, educator, and trauma professional David Treleaven notes that "mindfulness meditation can exacerbate symptoms of traumatic stress. Instructed to pay close, sustained attention to their inner world, people struggling with trauma can experience flashbacks, dysregulation, or dissociation."[21] This is a somewhat common occurrence in those of us living with post-traumatic stress disorder or a history of unhealed trauma, or if we're simply too swamped to make time for our emotions, allowing them to pile up and overwhelm our nervous system. When we busy ourselves, we often don't feel our feelings directly in the moment. In fact, we are rewarded for being able to compartmentalize our feelings and act like robots to get the job done. We move from one task to the next until the day is over, and we are ready to unplug. We stuff our emotions, only to be flooded by them later.

Being well versed in stuffing feelings, the flooding and overwhelm that come later are best prevented by allowing a little in at a time. Because I had trouble maintaining a regular conventional meditation practice, I decided to create one that works for me by combining the "Being with Uncomfortable Feelings" practice on pages 67–68 with the "Letting Go" practice on pages 155–56. I start by letting go of the tension in my body, my ruminating thoughts, and my expectations about the way the world ought to be, which carves out space for me to feel my feelings. I open to the feelings I have stuffed away throughout my day (or life), name and feel them, and allow for an embodied release. This process is helped by titrating my heavy and painful feelings, or allowing small amounts of the feeling to come through, so I don't get overwhelmed by them. I call my practice "Letting My Feelings Catch Up with Me." This practice is just one piece of my healing journey. While it alone doesn't heal my trauma, it opens more spaciousness in me so I can better show up in my life.

EXERCISE: TITRATING YOUR FEELINGS

As with every other exercise we offer, this is just a suggestion. If something becomes too uncomfortable, you are welcome to do something different or back out of the exercise until you are better resourced. If you are new to letting your feelings catch up to you, or you have experienced unhealed trauma, you can easily become overwhelmed by your feelings. This does not mean that you should not take the time to feel them, but it might be helpful to work with a healer and to practice feeling your feelings slowly to avoid overwhelming your nervous system and retraumatization.

Carve out space, silence, and stillness in and around you. Check in with your body. Is your heart beating fast? Are your shoulders raised? What is the temperature of your skin? Are your eyebrows furrowed? Is your jaw clenched? Begin letting go of the tension in your physical body. Relax your shoulders, unfurrow your brow, and unclench your jaw. Take three deep breaths with long exhales.

Begin to search around inside. Is there a feeling asking to be noticed? Is there a feeling you had to lock away because it wasn't safe to feel it before? Notice the feeling and open to it. We will use grief as an example of a feeling to be titrated, but this method can be used for any heavy or painful feeling.

Imagine that the entirety of your grief is contained in a funnel.[22] Near the bottom of the funnel is a valve that controls the flow of the feeling that is released and experienced. We can keep the valve off, and no grief will get released or processed. Or we can turn it just a little bit, so we can become familiar with that feeling. If the feeling we are exploring is grief, turn the valve slightly to release a bit of grief, and see if you can identify where in your body you feel a sensation. Is it a tightness in your chest? Do you feel anything in your throat or face? Did the temperature or sensation change anywhere in your body? If it becomes overwhelming, do not push yourself into feeling more grief. Close the valve. Call on your resources and come back to a sense of okayness. Be patient with yourself as you become aware of what it feels like to experience grief and begin to process it. You are building your distress tolerance.

If this practice is not overwhelming, titrate some more grief and then close the valve. Explore the feeling once again in your soma. Practice this exercise for a few minutes or until you feel like you have moved the locked-away feeling. As with all the exercises focused on feeling your feelings, remember to take deep breaths with long exhales and listen to your body to see if you can identify what type of release your body wants. Maybe it is jumping up and down, shaking your body out, crying, going for a walk, or a long, deep scream. Perhaps it is finding a safe person to connect with and share your experience.

Noticing repeated sensations in your body will help you become more familiar with particular feelings. You can, for example, discern that grief is present when you sense that your chest is tight and your face feels warm. The next time grief appears, you can be less overwhelmed by its presence and instead see grief as a friend who has come to visit, to share an experience, but is not welcome to stay too long. Similarly, rather than shy away from feelings of despair or fear because they hurt or are disorienting, you can acknowledge that they are there. Feel them. Even if it feels gross or like it might collapse you, practice being with the feeling for a short time. Remember from Step 1 that this experience, if you allow it to move through you, will only last between ninety seconds and twenty minutes. You can say to yourself, "I am feeling this feeling and letting it go now."

EXERCISE: 4, 7, 8 BREATHING

This conscious breathing pattern was developed by Dr. Andrew Weil and helps to reduce anxiety and bring us back into the present moment. Breathing in this way brings fresh oxygen to our organs and tissues. It was first taught to us by a GGN member, Josh, in circle. Since then, it has been a go-to exercise for LaUra when she cannot sleep or is stuck in an anxiety spiral.

In a comfortable seated position or lying on your back, exhale all the air from your lungs with a whoosh sound. With your mouth closed, inhale for a count of 4 through your nose. Hold your breath for a count of 7. Exhale through your mouth for a count of 8, making a whoosh sound again. Repeat this exercise for a few minutes at a time: inhale through the nose for 4, hold for 7, exhale through the mouth for 8.

Give yourself a moment to come back to your normal breathing before standing as the breathing exercise may make you dizzy.

Step 8

GRIEVE THE HARM I HAVE CAUSED

> Don't let anybody, anybody convince you this is the way the world is and therefore must be. It must be the way it ought to be.
>
> —Toni Morrison, *The Source of Self-Regard: Selected Essays, Speeches, and Meditations*

Aimee

We are all harm doers. We are messy human beings navigating the world under a paradigm created out of oppression and exploitation and maintained through violence. As individuals, we are blamed for the decisions of big corporations and a corrupt political order, as if our individual footprint can overturn the destructive behemoth that has been set in motion.

In a recent hospital visit, the staff wouldn't let me use a reusable water bottle. Instead, the hospital administered bottle after bottle of Nestlé water.[1] By day two, I had already used about ten bottles. In my mind, I could hear LaUra's voice on repeat telling me all about how plastic

We would like to extend gratitude to Sarah Jornsay-Silverberg for helping us expand this step as we moved it from the second step in GGN's 10-Step program to the eighth.

recycling is a sham, and plastic water bottles leach toxins into the water.[2] Plastic pervades every aspect of our lives. We destroy ecosystems to dig fossil fuels from below the Earth's surface and transform them into single-use plastics, only to ship them "away" to sit in landfills for hundreds of years or release them into our waterways, poisoning our aquatic cousins and drinking water. Plastic is in every ecosystem on Earth—in the food we eat, in giant patches in the oceans, and even in the deepest tissues of our bodies. Plastic has weaseled its way down to our geologic record.

Then I thought of the toxic-water situation in Flint, Michigan, a city about one hundred miles away from my hometown. For more than five years, drinking water from plastic bottles has been safer than drinking from their taps. Worldwide, nearly two billion people are without access to clean drinking water *right now*. In these situations, bottled water saves lives. My thoughts flip to plastic silverware, Styrofoam, diapers, maxi pads, straws, and all the useless packaging that haunts me. Sitting in the hospital, crinkling my plastic bottle between my hands, I sat with the complexity of this situation. For many, plastics are conveniences that come at the expense of energy, ecosystems, and health. For others, they are a lifeline. I felt powerless by this level of complexity. Should I not drink bottled water? Should I not use any plastic? What was the right thing to do? Was there a right thing?

In a paradigm full of *impossible choices*, no one can do the right thing all the time. We must constantly weigh our options, factoring in our needs, the available conveniences, our values, and the level of harm each option creates. This can make us feel shame, guilt, grief, frustration, or other heavy feelings. The dance of impossible choices is exhausting. And when we see others acting in ways that we don't agree with (using more water than us, flying, consuming meat or dairy), we might have the urge to judge them. This is where showing ourselves compassion and, once again, extending that compassion outward toward each other is necessary. Judging others for their actions while living in a broken system is a form of moral superiority that only serves to further divide and isolate us.

This is what our power-over systems do: blame us as individuals for toxic systems. Cultural messaging tells us that it is up to each of us to measure our carbon footprints, conserve water, use canvas shopping bags, decline plastic straws, and move toward a zero-waste lifestyle and that, until we have reached perfection, we have no right to critique systems. That notion is convenient for corporations that are gobbling up our planet, our resources, and our souls. We are all allowed to question destructive systems and, as we are able, withdraw our participation in them and hold them accountable for their grievous wrongdoings. Let us also stop policing the actions of everyone around us. Each of us has permission to be an imperfect change maker. Not only that, we must embrace our imperfections as change makers.

We are all doing harm to our planet and sometimes to each other and ourselves if we live in a paradigm that profits from harm and exploitation. Maybe we consume factory-farmed meat or dairy. Perhaps we drive a gasoline-powered car or fly on airplanes. It's likely that a number of us participate in fast fashion, buying clothes made in faraway sweatshops. The level of harm we participate in varies considerably from person to person and community to community. Unpacking our complicity is an endless task. And we must start reckoning with our levels of harm so that we can move closer to our values and undo these toxic systems that exist in the shadows of our shame. Our inability to be compassionately accountable to ourselves with regard to our participation in these systems helps keep the dominant culture afloat.

Sometimes, we make mistakes, play out norms that we are socialized into, act out of ignorance, or otherwise act out, and do harm that we could have avoided. We all make choices that are out of alignment with our values on occasion, and so many of us benefit from the complex system of privilege, power, convenience, and profit. We live in a world full of impossible choices, where we cause harm just by existing in this time and place. When this happens, we are invited to slow down, take a pause, and feel the grief that results from the harm we have caused. This process opens us to remorse, where we can identify ways to minimize harm in the future. Take it from Good Griever Alli Harbertson, who hosted our very first step program in Salt Lake City, Utah. She said,

> I felt literally lighter after acknowledging my complicity in these problems because it was such a relief to finally admit to myself and have others witness how much guilt and shame I felt to be in lockstep with a capitalist system that relentlessly extracts from and monetizes my life and the lives of basically everything I hold dear. I had never given voice to the shame of having so much plenty, privilege, and convenience at the expense of the living world and those who are oppressed. Looking back on this now, I realize this was ground zero in the process of decolonizing my mind to be able to accurately see the innate destructiveness of our collective Western worldview.

This step in our program is about acknowledging the harm we have done to ourselves, other beings, and our planet and creating space to grieve that damage. It is about examining the systemic power dynamics and our complex roles as harm doers, showing ourselves and others compassion for our mistakes and looking for openings to live in alignment with our values and minimizing further harm caused by our decisions and actions. This step is also about creating room for remorse, repair, and healing so that we have a chance of doing things differently.

EXERCISE: GRAPPLING WITH OUR IMPOSSIBLE CHOICES

Think of a time you made an impossible choice—big or small. We may have a million instances to choose from for this exercise. If you are new to feeling your whole range of feelings or holding yourself accountable, it is best to start small and work up to bigger instances of harm doing. Below are some small examples you may identify with:

- Maybe you were at the supermarket and forgot your reusable bags, so you had to use plastic ones.
- Perhaps you wanted to buy some fresh fruit for your family, but the only type available is grown out of season and shipped from a continent away.
- Maybe you live in a rural location, and shopping on Amazon is the only way you have access to some products that you need.
- Perhaps you have incorporated a meatless diet into your life and are visiting your Midwestern family who has made their staple meat-and-potatoes meal. You chose to share a meal with them.
- Maybe you were in a rush and chose something for convenience, like single-use plastics on a road trip.
- Perhaps you cannot afford to buy the organic, local, free-trade options and shop at a discount box store that imports material goods from sweatshops overseas.

In whatever example you choose to explore, reflect on what happened during and after the event. What were the conditions under which the impossible decision was made? Were you short on time? Did you act out of ignorance? Was a more sustainable or ethical option out of your budget? There is room for it all. Can you identify any feelings that arose during that event and can you feel it now? Is there a bodily sensation related to the feeling? Can you locate where in your body you feel it and describe the sensation. Maybe you feel a twisting in your stomach, or you notice an increased heart rate when you recall this situation. It is OK if you have trouble identifying a body sensation at this time. Just notice.

Take three long-exhale breaths and shake your body out for thirty seconds. Let the feeling go with self-compassion and an acknowledgment that you did the best you could at that moment. Perfection is not expected; nor is it required.

Remember this exercise and practice it again and again as you are faced with impossible choices. Remember to feel the feeling, name the feeling, and let it go with compassion and care.

Slow Down—Grief Is Calling

We are told that the things we buy and the energy we personally consume are at the crux of the climate crisis. Of course, these issues are part of it. But what we are facing is so much bigger than the neoliberal sentiment that we as individuals can solve this massive complex predicament. We must also grapple with the inherent racism, Othering, and the radical wound of separation from the more-than-human world that got us into this complex predicament in the first place. This is where systemic racism meets the environmental movement. This is where the impact of colonization, imperialism, and the stronghold of power-over structures comes in, necessitating grief work, remorse, and systems change.

The dominant culture and its toxic systems force us to make decisions that are out of alignment with our values. If we're not actively doing our inner work, unprocessed feelings resulting from these choices can accumulate in our somas. Feelings are complex and overlapping and are experienced in several ways. Remember, emotions are data points. However, in our time running support groups, we have noticed that if we dig deep enough, feelings of shame, guilt, or rage are often masking one of the most overwhelming feelings—grief. Many of us feel profound sadness about the state of the world and the destructiveness of our systems and helplessness about our involvement in said systems. Most of us are doing our best to survive and care for ourselves and our families by making impossible choices in an oppressive, exploitative paradigm. By our very nature, most of us do not want to cause harm. Nonetheless, to survive, we must. To avoid becoming cold, hard, or constricted by feeling like we are always betraying our values to survive in an unjust paradigm, we have to slow down to etch out time and space to catch up with our feelings.

Slowing down allows us to feel the weight of the harm we have caused. Grief doesn't operate at the dominant culture's speed. Grieving is a slow, deliberate process that is sticky and dense. Francis Weller calls for geologic pace when attending to our grief;[3] it requires time to hollow us out, readying us for reorientation. This process, if we allow it, can transform us from the inside out, but it will not work if we rush it. Similarly, we cannot repress, avoid, or try to bypass grieving. Grief work excavates our beliefs and expectations about our lives and the larger world, offering them up for evaluation. Grief work provides an opportunity to examine our past choices and assess how we want to live our lives moving forward. Grief work provides an opening.

As we enter the grieving space, we are asked to take accountability for our actions that have caused harm, invite remorse in, explore other pathways that minimize suffering, and let go with self-compassion. Remorse includes facing our level of complicity. Using compassion as a practice, we can remind ourselves that our intention was not to cause harm and that we are brave enough to witness the consequences of our actions. Remembering that grief is a slow process, we stay with it. In our slowness, we can reflect: Could I have done something differently? And, how will I do it differently from now on?[4] Roshi Joan Sutherland offers, "Grief is how we love in the face of loss, remorse is how we love when we've caused harm."[5] This is the path toward collective liberation, toward more inclusive paradigms that view all beings on Earth as valuable and worthy of dignity, safety, and belonging.

These new paths will not open to us if we remain distanced from the consequences of our actions, if we refuse to feel the pain of living in a destructive culture. It is in our nature to avoid discomfort, especially if we are not practiced in being with our painful feelings. Emotional processing work and completing the stress response cycle are initially uncomfortable, and if we have not learned how to be with our feelings, it seems easier to avoid them. The distance from our own feelings and our inability (or unwillingness) to process them creates an intolerance of other peoples' feelings and pain. This cycle causes more harm.

Those most callused to suffering are usually those winning the BAU game, which is only played at the expense of others. The fast-paced, hectic life of the modern world is addictive, providing great distractions and excuses for living in our dysregulation and disconnection. We stay in our familiar discomfort, avoiding the unfamiliar pain of processing our heavy feelings. Because many of us, especially those most invested in BAU, do not have much tolerance for distress, we look to the next thing that will keep us up, up, up. This leads to more suffering. It is a cycle that is tough to break out of, but it is absolutely necessary to do so.

Most of the people who are making the decisions about how much the world burns are addicted to this fast-paced competitive culture and are removed from feeling the consequences of their actions. They create distance from the grieving process with their success, money, and fame. Corporate CEOs and politicians are socialized to prioritize power and wealth at the expense of cooperation and equity. Jeff Bezos is one of the richest people on Earth, worth billions of dollars, and still refuses to pay his employees a fair wage.[6] For the elite, there is not much incentive to slow down and reflect on their actions or feel the weight of their decisions. And as long as they are winning at the BAU game, they will not.

The hard work of slowing down, grieving, and accessing remorse goes against the larger cultural norms. Healing and transformation never come from a frenzied, competitive place. Healing and transformation involve a willingness to slow down, bear witness to the consequences of our actions, exercise great compassion and change our behavior.

The Complexity of Privilege

> There is no amount of resilience that will sufficiently resource you to thrive in systems of adversity, inequity, and oppression.
>
> —Nkem Ndefo

Our roles as harm doers and harm receivers are complex. Each one of us has been on the causing and receiving end of harm. Depending on the bodies and situations we are born into, some of us benefit from these power-over systems more than others, and some of us have received more harm than others. Race, class, gender, sexual orientation, and other marginalizations all affect our roles as harm doers and harm receivers. But no one is inherently bad for existing in their body in this time and place. We are all a part of this complicated dance, and it is time to reckon with our varied levels of complicity and begin to heal our disconnections.

AN INVITATION TO STAY WITH IT

You are welcome to arrive in this moment, fully, with whatever levels of privilege and power you hold and wherever you are on your journey. If you have not already begun doing this, this is your opportunity to explore your levels of privilege and power with an open mind and heart.

This work is gritty and painful and absolutely indispensable for us to start moving toward the creation of truly just and life-centered paradigms. As we start to glimpse into our power and privilege, we can become overwhelmed by uncomfortable feelings urging us to turn away and never return to these pages. So, throughout this section, please pause anytime a break is needed. Notice when your triggers arise and explore the reactions within your body. Take long-exhale breaths and allow space for an embodied release (shaking, jumping, stretching, running, etc.). Connect to others who are following a similar path. Have heart-centered discussions with them. Show yourself patience and compassion as you dig into this work. Shame, however, is not helpful in this terrain. It isolates, divides, and ostracizes. It perpetuates the power-over structure, puts the blame on individuals, and moves us further away from liberation. We are all worthy of love and belonging, regardless of our level of complicity. Stick with this work. We need to lean into it to transform ourselves and the dominant paradigm.

Come back to these reminders as is necessary.

Those of us born in the dominant paradigm were raised in a culture created on the exploitation of human beings and of the more-than-human world. Each and every one of us is a harm doer just by participating in our culture. Once again, our level of complicity varies considerably, and there are infinite ways to cause harm in a system built on exploitation and commodification. Striving to live in relationship with life and life-supporting systems can help us minimize the damage we do, but unless we examine the ways in which the paradigm lives within us, we will perpetuate harm, oftentimes without knowing we are doing it.

Understanding privilege is not as simple as skin color, gender, wealth, or able-bodiedness. Each one of us has a complex and unique blend on a spectrum between privilege and marginalization. LaUra and I have the privilege of being well-educated and white-bodied, but as queer women living under the US poverty line, we also experience a mix of marginalization. It is often difficult for us to access health insurance and the health care we need. At the same time, we have more wealth than a vast majority of the global population, with access to running water, indoor plumbing, household warming and cooling, and safe shelter. We cannot afford to buy a home and, at the time of writing this, sometimes cannot afford to pay our bills, but we live in an area that is currently relatively safe from the impacts of the climate crisis. The list of complexities could go on and on.

Good Griever Sarah Stoeckl shared the following:

> "Grieving the harm I have caused" provided an entry point to reckon with my entanglement in systems and structures I did not create and yet, nevertheless, benefit from and perpetuate. These structures damage bodies, break hearts, steal land, tear open and poison the world, and also obfuscate their effects. And yet their effects are knowable and once realized cannot be unseen. This step is a call to accountability but an antidote to guilt and shame. It invites us into the circle of grieving rather than shutting us out, cloaked in judgment and imperfection.
>
> An imperative. An invitation. A permission. *Grieve the harm I have caused.*

EXERCISE: EXPLORING PRIVILEGE

Observe the Wheel of Power/Privilege below. This wheel is a starting point for understanding the complexity of privilege. It is not comprehensive, but it is designed to get you thinking and feeling about some of the privileges that exist for you. Where do you fall on each spoke of the wheel? What privileges do you notice that you had not thought about before? Is the wheel missing any spoke that you would add? Are there areas of marginalization you experience that you had not named before? How does your position of power, privilege, or marginalization make it easier or harder for you to complete different tasks (e.g., paying your bills, getting a job, making new friends, getting shot by police, getting into college, walking safely at night, or receiving medical services)? How does examining these aspects of your life make you feel?

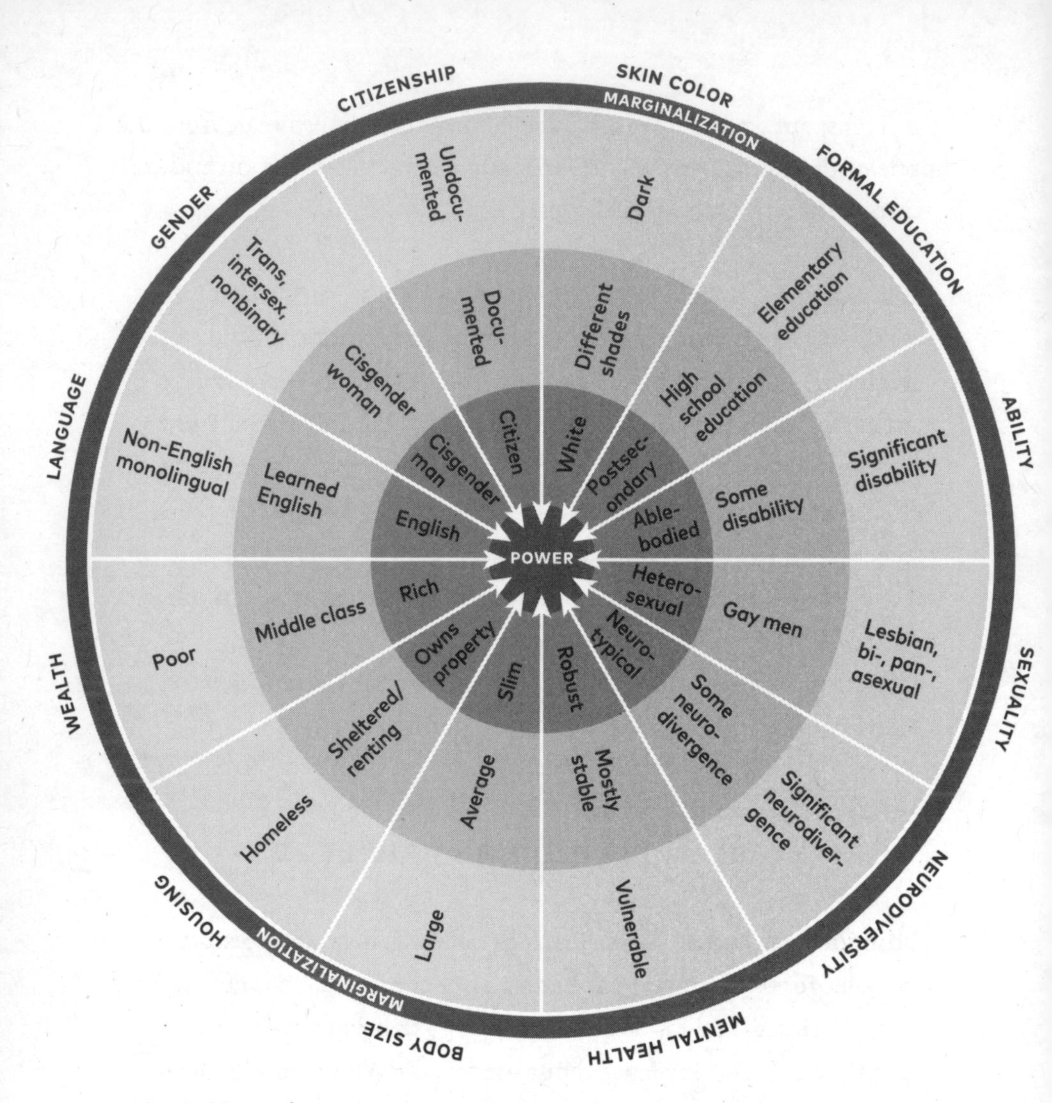

Wheel of Power/Privilege. Power and privilege are concentrated in the center of the diagram; marginalization and vulnerability increase with distance from the center. (Illustration adapted with permission from Sylvia Duckworth, @sylviaduckworth.)

Imagine a friend or family member. Map out your understanding of their power, privilege, and marginalization. How does their position of power, privilege, or marginalization make it easier or harder for them to complete different tasks? Does anything surprise you about this?

Now imagine someone you may perceive as Other. Map out your understanding of their power, privilege, and marginalization. How does their position of power, privilege, or marginalization make it easier or harder for them to complete different tasks? Does anything surprise you about this?

Take a few minutes to write down your answers to these questions and any reflections you have on this exercise. Pause to observe any feelings that arise in your body throughout this exercise. Do you notice any sensations? If so, where are they located? What quality do they have? Breathe through any sensation with long, deep exhales and shake your body out afterward.

Our level of privilege directly relates to complicity potential. For example, those with the most money make a significantly deeper destructive impact on our planet than those who have little money. All over the world, we see the wealthy consume more, buy up land (most of which was stolen from Indigenous people), and use more natural resources than those with low incomes. Meanwhile, BIPOC and poverty-stricken communities bear the brunt of environmental impacts from flooding, heat waves, exposure to toxic waste, air pollution, lead poisoning, and proximity to landfills. Overlooking our level of privilege and the amount of harm we are complicit in is a defense mechanism that we employ to protect our vision of ourselves and our position in the world. Exploring our role in the predicament is crucial, difficult, and absolutely necessary. Unless we slow down, explore our level of privilege, and feel the pain of our actions, we cannot reconnect and feel how our actions are harming other beings and our planet.

The Present Impacts of the Radical Wound

Trauma in a person, decontextualized over time, looks like personality.
Trauma in a family, decontextualized over time, looks like family traits.

Trauma in a people, decontextualized over time, looks like culture.

—Resmaa Menakem

It is critically important to link our radical wound to Othering and supremacy, which created the conditions for climate breakdown. As we tamed our living spaces and distanced ourselves psychologically, an ethos of human exceptionalism (human superiority over everything else) pervaded. Supported by anthropocentrism (viewing everything from a human perspective) and logocentrism (the focus on rationality and logic as the only way of knowing), a deeply ingrained, human-centric narcissism pervaded the dominant paradigm.[7] Not only have we prioritized humanity as the focal point of all of existence, but we have also further created hierarchies and separation within humankind, reserving access to resources for those most privileged.

We live in a socially constructed world of haves and have-nots. The haves are seen as justified for their power and wealth, and the have-nots are blamed for their lack of power and wealth. This division is predicated by the power-over system we first talked about in Step 1. As Staci K. Haines writes in *The Politics of Trauma*, "Power-over declares that it's okay to leave many in poverty, hurt and exploited, while we concentrate money, energy, power, and decision-making over others and the commons to very few. The social and economic distribution of dignity, safety, and belonging is how we construct who is seen as worthy of existing and who is considered expendable."[8]

We see this phenomenon play out largely through the practice of Othering, which is an evolutionary adaptive strategy meant to ensure our survival. Our ancestors' safety depended on identifying who was in their in-group and who was in the out-group. But this survival technique has been co-opted and manipulated in modern times to be expressed through current social norms. We know that race and gender, for example, are socially constructed and vary from culture to culture. We are not hardwired to see people of different races as Other, just as we are not hardwired to see gender nonconforming individuals as Other. The current hierarchies within the dominant culture are socially created by a set of privileged folks determining who is worthy and who is not. Our cultural norms perpetuate these ideas, and we follow along. It is all part of the consensus reality that needs dismantling. We can—and must—undo this social programming if we want to radically reconnect and transition to life-centered ways of being.

We are capable of extending the boundaries of our in-groups and of transitioning to a *power-with* model. In this way of being, hierarchy is removed, and power is shared. Reciprocity, caring for one another, and collaboration must be valued and prioritized. To get there, we must redefine who is worthy of safety, belonging, dignity, and resources and extend them not only to all people but to the nonhuman world as well.

THE LEGACY OF COLONIZATION

We cannot talk about the climate emergency without more deeply exploring the power-over system. Colonialism, a textbook tool of power over, has harmed all of us and, depending on our background, to different degrees. This tool lives on through us until we stop it.

We look to the wisdom of Resmaa Menakem for a road map to heal the disembodiment caused by racial injustice.[9] Menakem explains that trauma caused by Othering is alive in each of us, whether we have bodies of culture or are white-bodied.[10] For example, Menakem teaches that the Middle Ages in Europe was a time of "a thousand years of elite white bodies destroying less elite white bodies."[11] Public dismemberment, the Crusades, rape, land theft, and genocide were all happening to and around the white bodies during this time. Many years passed before the Europeans began leaving their homeland to colonize the Americas, but they brought their unhealed legacy of trauma along with them.

The unhealed wounding present in the colonizers' bodies proliferated as they wrought more trauma on whomever they perceived as Other in the New World. The wounding continues today with our endless wars, resource devouring, and systemic injustices. While Menakem's teachings focus on the effects of racialized trauma, we see how this is inextricably linked with a wide range of trauma from colonization and, more generally, exploitation. White-bodied colonizers deeply traumatized many marginalized bodies: bodies of color, queer bodies, bodies with disabilities, neurodivergent bodies, and feminine bodies. Colonizers, who refused to see their fellow humans as equals, extended this trauma to the more-than-human world, from our fellow animals' bodies to now-extinct bodies to bodies of land, water, and air.

We all carry cultural messaging in our bodies that keep the dominant culture alive. If we want to stop the colonial and exploitative cycles of trauma, Othering, and harm, we must begin to acknowledge, process, and heal the traumas that live in our own bodies from earlier generations.[12] This does not only apply to marginalized groups who are historically harm receivers. Healing colonial trauma is also vital for those of us who are the most privileged and whose white-bodied ancestors historically perpetuated harm, not only on those who they perceived as Other but on one another. As we move toward healing our traumas, we also heal the disconnection created within ourselves, each other, and the more-than-human world.

EXERCISE: JUST LIKE ME[13]

Slow down. Create some space in and around you. Take a few deep breaths with long exhales to arrive here, now. Bring someone you love into your mind's eye. Think about the characteristics of that person. Repeat the following words as you imagine this person:

- This person has feelings and intentions, just like me.
- This person has a story, just like me.

- This person is trying their best to survive, even if I do not understand the choices they make, just like me.
- This person experiences deep sadness or rage sometimes, just like me.
- This person experiences profound joy or meaning sometimes, just like me.
- This person has hurt someone else, just like me.
- This person has helped someone else, just like me.
- This person has needs, just like me.
- This person is yearning for safety, belonging, and dignity, just like me.

Now, transition to thoughts of loving-kindness toward this person:

- I wish this person safety, belonging, and dignity.
- I wish for this person to be free of harm.
- I wish for this person to experience peace and compassion.
- I wish for this person to find joy.

Is there anything else that you would like to wish for this person?

Create a moment of silent reflection on what that experience was like for you.

Next, think of someone who is harder to relate to right now. Maybe you are in conflict with a family member, or perhaps you can envision someone you perceive as Other. Try this exercise again with that person in your mind's eye.

After running through the exercise, take a moment to compare the two experiences. If it is helpful, record your reflections in your journal. What was it like doing this exercise with a friend versus someone you might be holding hostility toward? What similarities do you notice? What were the differences? What did you feel in your body when doing this exercise? What is it like to imagine that someone's struggles are similar to your own?

Can you extend this practice regularly when you find someone hard to relate to?

DECOLONIZE YOUR SOMA

We live in a paradigm built on coloniality, so we must decolonize our somas by healing our personal and intergenerational trauma. This is unlearning, embodied work. Our decisions and what we have access to in terms of opportunities and resources are all shaped by our presence and participation in these systems. To decolonize our somas, we must learn how these beliefs and norms live in our bodies and heal them from the inside out. We notice how our somas respond when we are perpetuating power-over sentiments. We discover where we have cast our boundaries in determining who is seen as Other and start challenging those boundaries. This work allows us to stop the chain of pain that we have inherited. Instead of responding like colonizers—out of domination, fear, and disconnection—we can bring ourselves to challenges as nimble, openhearted individuals. Strong communities are fostered through connection rather than Othering, privileging, and domination, but we must do the work to make this happen.

This means we must break down binaries and accept more complexity. Subjects like the afterlife, spirit and soul, beauty, consciousness, and how life started on Earth remain big question marks in this dualistic, reductionistic way of knowing. Decolonizing our somas calls us to embrace diversity, complexity, and what we have previously considered separate from us. Unlearning carves out space to consider that things might be different from how we perceive them and an openness to the unknowable. It is an ongoing invitation to be with the gray areas that constitute life without rushing to categorize or resolve them. We start to discover that rigid binaries such as "good" and "bad" mask complicated dynamics playing out roles of oppressor, oppressed, and complicit. Breaking down binaries opens us to the unknown, the uncertain, and, occasionally, the frightening. If we can do this with authenticity and courage, we have the opportunity to connect in community. And a genuine sense of belonging makes everything more manageable.

Decolonization is more than an intellectual exercise. Learning about colonialism and its ongoing impacts is important work, but alone, it is insufficient. Self-reflection and talk therapy, though they are a good start, are not enough for an individual to heal the legacy of trauma from collective wounds caused by imperialism and colonialism. Likewise, as we covered in earlier steps, such holding actions as strategy, protests, legislation, defunding, and reparations are also important but not enough to heal a community's trauma or to prevent the worst impacts of the climate crisis. We need a widespread shift in behavior and worldview, and that begins with the individual and our somas.

Resmaa Menakem's powerful somatic abolitionism program helps people heal intergenerational/racialized trauma. He suggests starting this work in affinity groups because (1) we cannot heal collective trauma alone and (2) sometimes the types of oppression between groups keep us from going deeper into the healing that needs to be done. We can establish all types of affinity groups for communities of practice, spaces that are able to uncover wounding that has been wrought on a specific demographic. However, we cannot stay in our affinity groups. We must be willing to encounter, embrace, and understand different perspectives from a variety of people to grow and create new ways of being.

This is uncomfortable work. And because we have been building our tolerance for distress, we are prepared to endure the discomfort of healing from racism, Othering, and coloniality. This process is helped by pausing and, once again, not rushing to solutions just to avoid the discomfort. We must practice an embodied noticing of our feelings and sensations by regularly checking in with our body. We can detect temperature and movements like our heart beat, our breath, and sites of tingling or pressure.[14] Noticing the quality of our felt experience allows us to begin to process it. As we practice this, our distress tolerance grows. When our uncomfortable feelings return, we recognize and lean into them instead of avoiding or stuffing.

This process does not happen overnight but is built up with practice. Menakem teaches that just like we go to the gym to build muscle, we must practice building our distress tolerance through repetition.[15] The more we are able to stay with our discomfort and practice the pause in times of uncertainty or distress, the more we will build our tolerance. We have to lean in again and again, in an embodied way, by paying attention to our sensations, pausing, and processing.

When we look deeply at the extent of the harm we as a culture have done to our planet, our bodies often constrict. Topics like privilege, race, and climate may make us feel shame, guilt, fear, or other feelings that create uncomfortable reactions in our somas. By exploring our feelings through pausing, we are invited to notice what happens to our bodies when we are out of our window of tolerance. Perhaps we get small, shut down, or venture inward in the face of such overwhelming issues. We can be with the constriction in our body, honoring our embodied strategy that helps us survive these painful, disruptive times. Without trying to fix our feelings or sensations, we can breathe through them, sending compassion to those places of constriction, as we discussed in previous steps.

If we are not practiced at sitting with our feelings and have unhealed trauma, it is possible that our survival responses will become engaged, leading to chronic dysregulation. In these instances, we are provided an opportunity to learn how to calm these responses by engaging in practices that bring us into the present moment and reminding ourselves that we are relatively safe. The calling of this complex and painful time on Earth is to be able to stay with the difficult work we need to do. However, if feelings and sensations work is keeping us out of our window of tolerance, it might be beneficial to work with a healer.

If we live in chronic dysregulation and constriction, we will cause more harm. Our survival responses, over time, become trauma responses and make decisions without our conscious awareness as an attempt to protect our safety. These trauma responses are conditioned responses from a previous time we encountered an unsafe event. Our body remembers this response and engages it again. Our conditioned responses are not always the wisest decision makers. As we covered in Step 2, when our survival responses are engaged, our brain becomes unable to process complex information. Our social engagement system is shut down, stripping us of our full ability to connect with other people and with the more-than-human world. Our trauma responses act out of self-preservation, often leading us to lash out, perceive danger where there is none, or shut down until the perceived threat has passed. By learning to recognize our conditioned responses, we have a chance at calming our nervous systems, coming back into regulation, and getting our social engagement system back online.

EXERCISE IN ORIENTING

Slow down. Arrive here fully. Settle into your body. Take a few deep breaths with long exhales. Begin to open to what is around you with all your senses.

Really look around and take in the shapes, textures, and colors around you. Notice and name five things you see. Maybe you see a houseplant, the pages of the book, a pen.

Deeply listen to all the sounds that you can hear, near and far. Is there something you are able to hear that you would have otherwise missed? Notice and name four things you hear. Maybe you hear the heating or cooling system in your room, the sound of an animal companion, a distant voice.

Bring your awareness to what your skin can feel. What does it feel like to notice where your skin intersects with the environment around you? Notice and name three things you feel. Maybe you feel the texture of the book, your fingertips rubbing against each other, your clothing touching your skin.

Really take in the scents around you. Can you smell something you have not noticed before? Notice and name two things you can smell. Maybe you smell lotion, deodorant, or coffee.

Explore your sense of taste. What can you notice about what is in your mouth that you have not noticed before? Notice and name one thing you taste. Perhaps you take a sip of water or tea to help you notice your sense of taste.

Take a few moments to reflect on this experience. Did anything change for you after this exercise? Do you notice any subtle changes? Do you feel calm or alert? Something else? Did you open to any sensations that you did not notice before?

Moving Away from Cancel Culture

In a world full of harm, how do we heal? How can we move toward repair and forgiveness of ourselves and other harm doers?

The answer starts with compassion and insight. Joanna Macy teaches the Shambhala Warrior Prophecy that was gifted to her by her teacher and friend Choegyal Rinpoche of the Tashi Jong community in northern India. She explains that compassion and insight are two required tools to help us navigate these tumultuous times as individuals and as a collective. We need the fuel of compassion, a force that reminds us to live with our whole hearts and not to be afraid of the suffering of our world. We also need insight, the wisdom that all of life is interwoven, and we all belong to each other. When we navigate the world using these tools, we are able to understand that the categories of good or bad are too simplistic, and we are all mixed up in a complicated dance of wounding, trauma, and survival responses. Joanna teaches, "The line between good and evil runs through the landscape of every human heart."[16]

Compassion and insight serve as reminders that no one is perfect and aspiring to be keeps us disconnected from one another. Instead, we can learn to embrace people as they are, with their gifts and their messiness. We must recognize that everyone has to start their healing journey somewhere. As we grow, we will mess up and ideally learn, make amends for our harm doing, and keep growing. The Adult Children of Alcoholics meetings, a space where LaUra finds significant healing, are closed by reminding participants, "After a while, you'll discover that though you may not like all of us, you'll love and accept us in a very special way—the same way we already love and accept you." We do not have to be best friends with everyone in our community, but we do have to respect each other's needs for safety, belonging, and dignity. This builds trust, and trust keeps communities together, so we can build cultures that are life centered. When we create spaces for authentic relating by allowing ourselves to be seen and seeing others as they are, we increase accountability and cooperation.

An emergent strategy principle that adrienne maree brown teaches is to "move at the speed of trust" by fostering critical connections over critical mass.[17] When trust and connectedness are the focus of the community, we start to form collective spaces and cultures based on reciprocity, nurturance, and accountability. This will be increasingly challenging if we are not committed to doing our inner work, especially as the impacts of the climate catastrophe bear down on us. As storms, droughts, fires, or pandemics threaten our safety, a dysregulated nervous system will engage our survival responses, contributing to Othering, self-protective actions, and violence. Our communities will then be led by shame, guilt, fear, rage, and other divisive feelings leading to mistrust and more harm.

The question then becomes, How can we build nurturing communities and cultures that create accountability when harm occurs? Because it will occur. In the past decade, the so-called cancel culture has emerged as one method for holding harm doers accountable, both in mainstream circles and in activist spaces. The threat of "canceling" each other or being ostracized from our communities is deeply destabilizing. Canceling also silences ideas and impedes the courage we need right now. If we believe there is a risk of being called out or removed from our community, we will not challenge the status quo. It leads to infighting and cuts down well-meaning but misguided members of our communities. Canceling is not effective for solving problems or creating strong bonds. This method for holding harm doers (or perceived harm doers) accountable is often destructive. Instead, when harm is caused, adrienne maree brown suggests, "We will respond not with rejection, exile, or public shaming, but with clear naming of harm; education around intention, impact, and pattern breaking; satisfying apologies and consequences; new agreements and trustworthy boundaries; and lifelong healing resources for all involved."[18] Much needed covisioning, healing, and transformation can happen if we are not in fear of being kicked out of our community or that the love we receive is conditional. Belonging provides security. Security allows for courage, growth, and healing.

Cancel culture is bound up with our profoundly broken accountability methods like policing and the judicial system. There are more loving and just road maps for accountability in a community setting that center justice and repair not just punishing and extracting people from their communities. One such model is the transformative justice (TJ) movement, which provides hope that we can both transcend our deeply dysfunctional, racist, and ableist systems of punishment and move away from the downfalls of cancel culture. TJ is a method for establishing communities where we work toward "healing, accountability, resilience, and safety" for everyone.[19] A popular adage in healing spaces is "hurt people, hurt people." Through the TJ framework, fellow community members understand that violence is caused by someone who is experiencing deep pain. And behind that deep pain are probably deeply flawed social systems that failed that person or community. The writer, public speaker, and community educator Mia Mingus asks, "How can we respond to violence in ways that not only address the current incident of violence, but also help to transform the conditions that allowed for it to happen?"[20] The goal is to repair harm on all levels and allow relationships and trust to inform the way.

EXERCISE: TONGLEN MEDITATION FOR THE WORLD

Tonglen meditation, also known as "taking and sending," comes from the Buddhist tradition. This practice challenges our avoidance of suffering and helps us visualize and transform the pain of the world. On the in-breath, we visualize taking in the pain of the world. On the out-breath, we send out love and compassion. This is a great practice when you feel overwhelmed by the suffering in the world.

To practice, clear some space in and around you. You can sit up straight or lie down. Take a few deep breaths with long-exhale breaths. Invite in openheartedness and spaciousness.

As you breathe in, visualize the pain of the world, the heavy feelings, the suffering. Imagine the texture of those feelings and the quality of the suffering. Maybe you want to focus on one aspect of the world in pain: the rivers, the oceans, the soil, or a particular being. Take in all the negativity of the world, and imagine your body is metabolizing it. Imagine you are bringing the pain into your heart, and your heart is strong enough to transform it.

As you breathe out, focus on sending love to that suffering and pain. Visualize a light bringing healing, warmth, and love to the world, overcoming the suffering. Send the rivers, the oceans, the soil, and a particular being love, warmth, and healing light.

Continue this practice for any part of the world or people who need healing. You can focus on an individual, a collective, or a whole ecosystem. There is no wrong way to do this practice. The basics are that you are breathing in the suffering and pain, allowing your body, breath, and healing light to transform it, and breathing out love and compassion. Consider adding this as a regular practice.

Step 9

SHOW UP

> Courage has a ripple effect. Every time we choose courage, we make everyone around us a little better and the world is a little braver. And our world could stand to be a little kinder and braver.
>
> —Brené Brown

Aimee

On more than a few occasions, I have seen how showing up can save lives—including my own. Before I was hospitalized for depression, showing up authentically was unthinkable for me. I nearly killed myself because asking for help for my depression did not seem like an option. The emptiness I was experiencing made me feel like a failure, an outsider. There was such a stigma around mental illness and depression that I thought no one would understand what I was going through. *I* didn't even understand what I was going through. How could I be so depressed when I had been given so much in life? Why couldn't I just be normal?

As the depression worsened, so did the suicidal ideation. Reaching out to family for help was out of the question—the shame I felt was too big. When I tried to share my feelings with my friends, most of them would tell me to cheer up or try to distract me. But no amount of distracting ever made the void go away. It was right there waiting for me, all day, every day, and especially at night.

LaUra was someone I could show up with. On one of those late nights when the depression was overwhelming my sense of reality, LaUra and I sat by the riverside in our college town. While listening to the water flow from the bank, something inside me cracked open. After two brief stints in a small psychiatric unit, I did not want to go back to the hospital—the intake process was grueling and dehumanizing, with hours of paperwork and questioning that ended each time in a strip search. I could not do it a third time, yet it felt intolerable to stay alive. The mixture of psychiatric medications was not helping me get better, and the side effects were only increasing my depressive symptoms. LaUra held my hand and told me she needed me to stay alive. She had already lost two best friends and did not want me to be next. She reminded me to hold on and that the meds needed some time to work and that I could call her at any time. By the way she said it, I trusted she meant it. We parted ways and went back to our respective apartments for the night.

Hours later, I sat on my bathroom floor with a kitchen knife. I needed to go to the hospital, but I didn't have the strength to go through the intake process again. Instead, I calculated the type and depth of cut that would have me bypass intake and wake up in the emergency room. When I saw the blood, I sat for a moment and then picked up the phone. Within minutes, LaUra was seated next to me on the bathroom floor, our knees touching. After seeing that my wound would not bleed out, she stayed with me the rest of the night. In the morning, LaUra drove me to a partial hospitalization program, where I was admitted for the third and final time for many years.

After I was discharged, I had to face the next part in my journey of depression: the humiliation of having to tell people that I had just been in the mental hospital, not just once but three times. I was surprised by many of the responses I received. One by one, friends and acquaintances began opening up to me. They told me stories about their cousin, friend, or parent who had also been to a psychiatric hospital. I found that once I started talking about my struggles with depression, others shared with me their journeys with mental illness. It was then that I made the firm decision to show up as my authentic self and tell my story, even when it felt too vulnerable. I knew that the truth, my truth, might help someone feel a little more courageous in their own journey.

This step in our program has a special place in my heart. LaUra showed up for me, and it saved my life. Now I have the honor of showing up for others. As a result of my willingness to be vulnerable, I have had loved ones call me in moments when they have struggled to hold on. It's hard to call anyone in times of absolute desperation, and they were courageous enough to call me. These days, I don't have to hide anymore—I know that if the illness overwhelms, I can be courageous enough to call one of them, too. Courage expands the more we bring it into the world. Acts of courage invite more courage, and these times require endless amounts of courage.

A Willingness to Get Vulnerable, Together

> What kinds of culture, knowledge and community structures would we be able to create if we could nurture one another without our armor on, if we could draw out and develop the gifts in one another, if we could care for another in concrete, meaningful ways, and could protect one another from systemic harms and forms of structural violence, even as we're struggling to dismantle them?
>
> —Nora Samaran, *Turn This World Inside Out: The Emergence of Nurturance Culture*

We must build communities based on vulnerability and openheartedness so that whether we are feeling on top of the world or alone at the bottom of a well, we can show up for ourselves and each other. Like many skills, we get better at it the more we practice it. Removing our armor and being together with open hearts is often easier said than done. Practice groups are essential to help us unlearn our cultural conditioning and trauma responses and relearn how to be with one another. Spaces must exist where we feel we can show up fully as our messy, vibrant, and vulnerable selves and be held collectively. Honest interactions develop trust between people, and trust strengthens healing bonds. What is more, our nervous system loves being embraced by our communities. When we authentically relate to other (safe) people, we can coregulate our feelings, leading to grounding in the present moment. This phenomenon is labeled *attunement*, a process that helps bring the social engagement system online. (In Step 2, we introduced the vagus nerve and how activation of our social engagement system is key for connecting to others.)

Sharing our authentic selves—the good, the messy, the painful—can feel risky. Brené Brown writes in *The Gifts of Imperfection*, "Heroism is often about putting our life on the line. Ordinary courage is about putting our *vulnerability* on the line."[1] When we open up wholeheartedly to others, we also open ourselves to their judgments and reactions. We may worry that we will not be accepted. Maybe we feel that our feelings are too intense, or we protect against being seen as crazy or overreacting. The dominant culture's stigma against mental health makes it difficult for many of us to be open about our internal anguish. As our collective stressors intensify, we might be overwhelmed by questions: *Why do I seem to be the only one feeling these feelings so deeply? Am I overreacting? Why aren't more people as alarmed as I am? Why is it so scary to open up and share about my experiences of these collective issues? Are there spaces where I can be real about the panic, grief, or anger I am feeling?* These are common thoughts that so many are experiencing right now. It is not surprising that many of us are reluctant to shed our armor and connect.

In stressful times, many of us unconsciously constrict when we could be connecting. This response uncovers a deep wound that needs healing. Maybe vulnerability was not modeled for us in a safe way by the adults in our early lives. Perhaps vulnerability was used against us, and we were taught to internalize our feelings and ignore our body sensations to prevent further wounding. When we are lacking safety, belonging, or dignity, our somas make the decision that it is in our best interest to armor up, disconnect from our bodies, and distrust others. However, we know from life experience that over time the armor becomes too heavy and leads to avoidance of the issues, physical ailments, increased intensity of uncomfortable emotions, and, in our case, eventually depression. If showing up has not been a safe experience, we may need to find a trusted healer to help work through our wounding and restore safety in vulnerability. Additionally, hiding our true selves out of fear of being hurt again denies us the benefits of belonging in a community.

As we heal our wounds, integrate our shadows, and feel our full range of feelings, we learn that to show up in our communities, we must also show up for ourselves. We aim to become more embodied, more connected to our senses and sensations, and more and more alive and then, from this place, reconnect outwardly to our communities.

Many of us do not have safe spaces to share vulnerably and authentically about the pain we feel over our individual experiences or to process our overwhelm about collective events. Messaging received from the dominant culture tells us to keep quiet, suffer in silence, and cope through launching into action to lessen our individual or collective pain. Good Grief participant Zeinab Benchakroun explained,

> For a long time, I have been holding an entangled ball of emotions related to the climate crisis in my chest. I was not even aware of it. I was in pain without being able to look at it. When I joined GGN, I realized that I was not alone in my suffering over the state of the world. I came to understand how much the modern culture represses this kind of conversation. The dominant ideology of growth and extraction has disconnected us from our hearts, from each other, and from the nonhuman world. GGN

taught me to be more comfortable talking about my pain and gave me tools to be able to process it in a healthy way. And the process, of course, is ongoing work.

The privatization of our pain plays out in many movement spaces, especially in our mainstream activist communities. Activists are underdogs working to intentionally dismantle the status quo. Because of this, those of us who are activists often face skepticism, defeat, and ridicule for working for a more just, life-centered world. We commonly receive messages to armor up and ready ourselves to battle for our cause—and vulnerability is weakness in battle. The armor is in place to protect from loss after loss as we fail to see our vision fulfilled in the world around us. But the refusal to be vulnerable only serves to alienate and disconnect, making us far less effective as change makers and more susceptible to burnout. It wears us down somatically. Francis Weller wrote, "We have to give up the idea of private salvation and the idea of private healing. That's all fantasy. We either heal communally or we don't."[2]

EXERCISE: PRACTICING VULNERABILITY

Brené Brown teaches that vulnerability is the key to so much of what we need in the dominant culture: connection, creativity, innovation, and change. She asserts that many of our issues can be resolved if we let ourselves be seen, deeply.[3] And if we hold space for others to be deeply seen. Yet, some of us may not know how to do this.

If you are yearning for connection or to change the way you are living your life, you do not have to wait for a support group to bring authentic connection into your daily life. You can start showing up in your life in wholehearted ways.[4] Contemplate the following questions:

- Do you find yourself avoiding hard conversations because you fear exposing too much of yourself?
- Do you answer honestly when someone asks how you are or do you brush off the question?
- If you have had a hard day, do you reach out to a loved one for support and encouragement?
- How often do you find yourself shying away, hiding, or avoiding so that you do not have to be vulnerable with those you love?
- If a project you are working on does not go the way you planned, how often do you share your disappointments with someone else?
- Are you longing for a deeper connection with your friends and/or family?
- Do you feel that you are worthy of connection and belonging?

Reflecting on these questions can help you identify situations in your life where you can begin leaning in, practicing courage, and showing up authentically. Even if vulnerability is not the precedent in your relationships, you can start practicing today. The more you bring in vulnerability with the people around you, the more people will feel comfortable reciprocating openness. Try it!

Begin looking for openings to practice vulnerability with those you see each day. Maybe you find a moment at the dinner table to discuss how the climate crisis makes you feel. Consider offering a listening ear to a friend who just lost a loved one. You could authentically answer how you are doing the next time someone takes the time to ask. Maybe you can have that hard conversation with your partner that you have been putting off about a repeated behavior they engage in that leaves you feeling unseen or underappreciated.

Be discerning with this exercise. Anytime we practice vulnerability, there is a risk of someone not responding the way we want them to, which can cause us pain. If practicing vulnerability is new, start small. Open yourself a little at a time and more often. This allows the authenticity and trust to grow and deepen.

Make some time to reflect on your vulnerability practice through meditation or in a journal. How did it feel in your body to be vulnerable in your daily life? Did you experience discomfort? If so, where? Were your muscles tense? Relaxed? How did the conversations go? As you continue repeating this practice, take notice of any ways your relationships have shifted or remained the same. Can you identify any patterns?

Showing Up for an Uncertain Future

> One of the most calming and powerful actions you can do to intervene in a stormy world is to stand up and show your soul. Struggling souls catch light from other souls who are fully lit and willing to show it.
>
> —Clarissa Pinkola Estés

Because the future is uncertain, we understand the importance of stepping into our personal power in the present moment: living each day to the fullest, pursuing meaning and joy, and focusing our attention and actions on increasing connection, growth, and healing and unlearning our cultural conditioning.

Each of us holds gifts to share with the community right now and throughout the Long Dark. The strength of these gifts exists in our ability to endure discomfort, do our inner work, and come together in community even when despair threatens to overwhelm. These gifts do not come from being perfect or having all the answers. Instead, our gifts become offerings when we embody our personal power and pursue what brings us alive. Our power is energized by our ability to embrace the grit and joy of living through these times. Even as those who have been overcome by despair tell us that our actions are hopeless, we refuse to stop planting seeds for the future.

Our ability to show up wholeheartedly inspires others to show up wholeheartedly. We show up authentically through the hard experiences and celebrate the delights. The heart-centered revolution involves awakening and aligning to our superpowers so that we can follow our values and plant those seeds and cultivate the conditions for life-sustaining paradigms to emerge.

FOLLOW YOUR YES

There is no one right way to show up in chaotic times. It is a constant discernment process to decide how to use this one wild and precious life.[5] We can show up for a friend by making time for a deep conversation. We can show up for our community by sharing the surplus food from our garden. Or we can show up for ourselves by turning off our phone, saying no to the next protest, and going on a hike with our dogs and partner. Sometimes, showing up for ourselves also means lying in bed, coloring, or watching a favorite TV show. However we choose to show up, what is important is that we are doing it for reasons that recharge us, not drain us.

Showing up out of obligation, because we are worried about what others will think of us if we do not, or because we are afraid our absence will hinder progress, has been celebrated in dominant culture and even in activist culture. After years of facilitating GGN circles, we have witnessed innumerable people who are burned out from showing up for reasons that drain them. When the complex global predicament is transforming our planet as we know it, we can feel like we must attend every march, boycott any products created by unethical corporations, keep current on every article and report, and argue with any distant cousin on Facebook who is grossly misinformed.

Some of these actions help pave the path for new ways of being. In the case of protests and events, showing up in numbers can make a powerful difference in the decisions our politicians make and serve to virtue signal, helping to bring other change makers aboard movement spaces. Being educated on what is happening around us can guide us in making more informed decisions. It is a tightrope walk that each of us must take because our mental and emotional resources are limited. Will reading that additional article or attending that march topple us over? By practicing the seven types of rest highlighted in Step 7, we can start to notice which types of activities provide energy and momentum and which ones drain us. The more we show up out of obligation or for superficial reasons, the more we deplete our valuable energy and capacity. And if we keep showing up at the expense of our need for boundaries, joy, and meaning, we will bottom out. We cannot pour from an empty cup.

Because we are inundated by crises, it can be difficult to notice the difference between actions that preserve the soul and ones that chip away at it. Our panic is screaming, "Emergency!" and that makes it awfully hard to hear the voice inside of us that is trying to gently illuminate a way. We cannot always tell which ways of showing up are fruitful for us—and there is no hard-and-fast rulebook with the answers. What recharges one person drains another.

How can we know when we are showing up for the right reason? Once we calm our nervous system and quiet the alarm bells, we can practice discernment. What is in alignment with our values and with our integrity? We take cues from our somas by turning inward in contemplation, looking to our intuition for guidance and noticing what we are pulled toward. Staci K. Haines teaches that a fundamental lesson in somatics is that "the body learns on 'yes.'"[6] She continues, "We organize ourselves, our somas *toward* something. We think, act, and practice toward a possibility. It is harder to place our attention on stopping something without attending to what we are going to practice instead. It is much easier to fully engage a healing or change process, especially through the difficult times, when we have a vision big enough to compel us forward."[7] Just as we discussed in Step 5, sensations in our bodies can help us notice if we are excited and energized by a particular action or if we are dreading showing up in a particular way. What are you being pulled toward? What are the yeses in your life?

The more we embody our full selves, without apology, and bring ourselves into relationship with the world around us, the stronger we will be as individuals and as a collective. But we do not get there out of the gate. It requires noticing, practice, enduring discomfort, and being with each other without fixing. It requires owning our stories, doing inner work, and healing our wounds. Once again, if we have unhealed trauma surrounding vulnerability and embodiment, it can complicate our ability to listen to our intuition, sensations, and somas. Vulnerability always carries risks, but the more we practice discernment, the more comfortable we are in our own skin, and the better we can show up for our yeses.

Cultivating Boundaries

> Boundaries are the distance at which I can love you and me simultaneously.
>
> —Prentis Hemphill

Even when we listen to our intuition, show up in ways that recharge us, and learn to follow our inner yeses, we can never predict or control others' reactions. And that is where the power of cultivating boundaries comes in. This may sound counterintuitive in a step that is all about showing up, but boundaries and vulnerability are two sides of the same coin. Knowing what recharges us versus what drains us—or even just what is OK and not OK—is a key part of connecting deeply with ourselves and others. We maintain strong boundaries to care for ourselves, to protect ourselves, and to restore ourselves. It helps us be more trusting and trustworthy. adrienne maree brown reminds us that our noes create space for our yeses to be heard. "Boundaries create the container within which your yes is authentic," she writes. "Being able to say no makes yes a choice."[8]

For many of us, it is difficult to say no, even under normal circumstances. Add in a burning planet, global social injustices, and a sixth mass extinction, and it becomes ten times as hard to say no to doomscrolling, organizing, or attending events. Some of us never learned appropriate boundaries, and this is especially true for women, who are often socially conditioned to be polite and accommodating and not to hurt others' feelings. We are often expected to give beyond our means, at the expense of our time, energy, and spirits. As we have discussed in previous steps, we know these truths all too well.

Oftentimes, we fail to draw lines when we need them most—particularly when we are affected by a crisis. Those of us who are directly affected by climate breakdown, whether by wildfires, flooding, droughts, storms, or other events, may want to rush to protect ourselves, our families, and our homes by finding quick solutions—to organize events, read more articles, call the media, or attend every single town hall. These struggles are long. We must carve out pockets of time within them to grieve, find our footing, connect with loved ones, and experience joy and meaning. We can trust others in our community to do some of the heavy lifting as we draw protective lines around ourselves to rest and heal. That way, we can show up wholeheartedly in both times of chaos and in times of joy.

Many of us do not know how to set and maintain boundaries. Perhaps communicating boundaries was not modeled for us. Or we may have grown up in an environment in which our boundaries were crossed frequently, so we do not recognize when someone is not respecting us. We all have boundaries, whether we are able to convey them or not. When we do not maintain our boundaries and clearly communicate them, they inevitably get crossed, and we are left feeling disrespected and resentful. We shut down and may not bring the situation up to the boundary breaker—and so the pattern repeats.

Setting boundaries can be extremely uncomfortable at first, especially if you have been conditioned to put others' feelings before your own. How will someone react when you tell them no? Will they think you do not like them? Will they think you are a mean person? Will they stop talking to you? These are all valid fears. Some people do not respond kindly to boundaries, but you also might be surprised to find that others welcome them with open arms. In fact, they might even be encouraged to follow your example. Ultimately, we cannot control how other people respond to our boundaries, but if we fail to communicate them, we will end up feeling stepped on either way.

We can begin cultivating boundaries by turning the inquiry inward. How do I practice boundaries? Are they helpful, or are they lingering conditioned responses that once kept me safe but now keep me from connecting? Rather than immediately and automatically saying yes to every request or invitation, we can pause and take the time to question what is being asked of us. We can ask, *What are the demands and expectations being placed on me and by whom? Others? Myself? Are those expectations realistic? Are they asking more of me than I have to give? Is it something I am excited about? Am I following my passions? Am I hearing my inner yes? Does the next step make me come more alive? Does it contribute to healing, growth, or repair work?*

The more we understand our boundaries and the requests being made of us, the more discerning we can be. There are times when it is appropriate to stretch our boundaries—say, for our best friend or our sister. We can safely ease our boundaries when someone has earned our time, care, and energy, but when someone expects or demands us to stretch, it can be damaging. Boundaries can be made porous with our intentions and discernment. Just like vulnerability, boundary setting is a practice. The more practiced we are in setting boundaries, the more time we can spend with people who accept and encourage us to be our authentic selves—even when we say no.

JOURNAL EXERCISE: BOUNDARY SETTING

We can start setting boundaries in the same way we practice vulnerability—by practicing in small increments with people who we feel are safe. The next time it feels right, try setting a boundary around something small—practice by saying no to an invite for coffee if you'd rather spend time with your partner or by skipping a weekly meeting if you need rest. After practicing this a few times, move to larger boundaries. Perhaps the next step is to tell a friend or family member that you will no longer tolerate a behavior of theirs that causes you harm. Maybe you choose to sit out the next protest and explore sensory rest instead. Or, maybe boundary setting looks like showing up for a limited amount of time. You are free to explore using intention and discernment where your boundaries are drawn. What moves you toward a yes and what has you recoiling in a no?

In your journal, reflect on your relationship with boundary setting. What does it feel like to set boundaries? Is it easier or harder with those you love? What has kept you, or still keeps you, from setting boundaries? What does it feel like in your body to set a boundary? Do others challenge you when you set a boundary for yourself? Are you setting boundaries to preserve yourself and your energy or to self-protect and remain constricted? What does it feel like when someone has pushed through your established boundaries? Do you have an embodied reaction when this happens? What are some ways you can prevent boundary intrusion in the future? Do your boundaries need to be stretched, or are they firm?

Talking, Trusting, and Feeling

> Do not ask what the world needs. Ask what makes you come alive, and go do it. Because what the world needs is people who have come alive.
>
> —Howard Thurman

LaUra

It's hard to let anyone in or to show up fully when you are surrounded by layers of abuse, neglect, and loss. My early life was dominated by the three rules of dysfunctional families: don't talk, don't trust, don't feel.[9] I spent years avoiding telling anyone about the harsh reality I was living in. Neither my friends nor my teachers knew, and my sisters and I didn't even talk among ourselves about the chaos we were living in.

When conversations got too intimate, I would make jokes or drive the conversation in another direction. On many occasions, I avoided my feelings by binge drinking. I didn't know how to live in a way that reflected my interior world. And the outer world around me gave me messages that my trauma was too much for most people.

When Aimee and I met, we saw a light in each other that was yearning to be seen. She was patient and loving and did not shy away from my pain. She did not try to fix it either. Aimee's stint in the psych unit was the wake-up call I needed to take my own issues seriously. If she could be so open about her mental illness, I could too. My story needed to be integrated into my lived reality. I needed to show up for myself and stop hiding my wounds, even if they were difficult for others to hear about. As I began being more open about my anxiety, depression, and trauma, a few of my friends distanced themselves from me. Even today, many people shut down or turn away when I share portions of my story, unsure of how to hold space for such deep wounding. Cultivating boundaries is necessary as I decide who to share my journey with.

As mentioned earlier, attending Adult Children of Alcoholics was life changing. These support groups provided a container that, for the first time, felt big enough to hold my pain. The experience was so profound that I carried many of these meeting norms over to the GGN spaces.

But despite years of healing work, I didn't know how to speak my truth publicly or be a change maker in a seriously wounded world. This all changed in graduate school, when I moved from Michigan to Salt Lake City, Utah, spending literally every last cent to get there to study with Terry Tempest Williams and the stellar lineup of professors, including Teresa Cohn and Brett Clark, at the University of Utah.

Terry's fierceness liberated something in me. Her ability to stay with the tension, to go with the flow, and to speak off script and from the heart modeled a trust in the moment and in the people in the room that I had not seen before. She lived and breathed adrienne maree brown's emergent strategy principle: "less prep, more presence."[10] This only worked because she insisted that we show up, hearts open, to each experience. I quickly learned that my normal tactics of sitting in the back of the classroom, avoiding engaging in class discussions, and still earning top grades would not fly in these spaces. These classes demanded heart-centered participation. For the first time in my life, I was encouraged to share my life experiences, surviving my traumas and losses, as strengths that forced me, early on, to start questioning the dominant culture. Terry, and her husband Brooke who often attended our classes, asked us time and time again, "What brings you alive? What is meaningful to each of you?" The experience was so raw and foreign to me that my anxiety went wild. Each class, I was being pushed to the edges of what was possible for me. Stinne Storm, a friend in my graduate cohort, called Williams's courses "graded group therapy," and it really felt that way sometimes. To live differently, we must explore our edges to see new horizons.

Throughout my studies, I came to an embodied understanding that the single most important thing I could do was to stop pretending, even for a moment, that these systems are in place to protect humanity or other life on this planet. By showing up authentically in these spaces, I learned the importance of community, of bearing witness to each other's stories, of trudging through the muck together even if we could not solution our way out of it.

No one heals alone. We heal in community. We inspire each other by our quality of showing up. But we must be willing to show up first. We can't shy away from our experiences for too long; otherwise, we submit to a small life. The world needs us to expand as everything around us constricts. Showing up is still a struggle for me. I'm continuously having to practice maintaining boundaries and being vulnerable. Not everyone deserves to hear our stories. But we do have to tell them. We get to be discerning about who we lower our armor for.

EXERCISE: LETTING YOURSELF BE SEEN

Choose another person you feel safe and comfortable with for this exercise. Go to a place where you will likely not be disturbed for this intimate experience. You can sit or stand, whatever is comfortable for you and your partner. Set a timer for two minutes. During this time, face each other, looking into each other's eyes without exchanging words. As you silently look into the eyes of your partner, you may feel an urge to laugh or cry. It may feel excruciatingly long or very connective. All of this is normal and OK. Do not distract yourself or leave your body. Stay with your partner and the moment. Be with what is unfolding.

After this experience, spend some time reflecting with your partner about the exercise. Take turns answering questions like, How did it feel to be seen, deeply? Did you notice any feelings, sensations, or urges arise? Were there any thought patterns that you can identify? In what ways was your partner's experience similar or dissimilar to yours?

Step 10

REINVEST IN MEANINGFUL EFFORTS

> Ours is not the task of fixing the entire world all at once, but of stretching out to mend the part of the world that is within our reach.
>
> —Clarissa Pinkola Estés

Change is not linear. Change is slow, then fast. It is frequently messy, sometimes backward, occasionally upside down, and often unrecognizable at first. As we develop a relationship with ongoing and increasing unpredictability, we need to cultivate inner worlds that are alive, responsive, and willing to remain open and connected. This is why most steps are pathways of transformation that begin with our internal programming—examining our survival responses, observing our relationship to reality, exploring our boundaries, unlearning, and leaning into the discomfort.

In this final step, we take what we have learned and venture outward once again into this chaotic, beautiful, broken world in crisis. Now that we have begun healing, practicing, opening, stretching, and getting disoriented, we can begin living the new ways of being. What is possible when we take the time to deconstruct our deeply flawed worldviews and build back up with individual and communal actions?

Heart-Centered Activism

Much of traditional activism focuses on top-down solutions. We try to solve our world's problems by demanding that leaders enact laws and regulations that trickle down to bring about justice, equity, and biodiversity conservation—in theory, anyway. Similarly, how many of us have been told to "vote with our dollar" and to change corporate decision-making through our consumerism? While these actions may result in slow, incremental change over time, these methods of engagement terribly limit our sense of personal agency by putting decision-making in the hands of someone else, usually either a CEO or lawmaker. As we have seen, too many politicians and corporate leaders have agendas that are profit focused instead of life centered. If we keep abdicating our personal agency to the political and corporate elite, we will see the same results we have been seeing—Earth growing hotter, the wounds of injustice festering, the rich getting richer, and more species being driven into extinction.

We founded GGN to flip top-down activism on its head and create a movement that comes from the bottom up—starting with each of us reclaiming our personal agency at a time when so many of us feel too small to make a difference. Taking action does not necessarily mean risking arrest or giving up on society to live in the woods. Meaningful actions exist in the decisions we make with compassion and love toward ourselves, others, and the more-than-human world. How can we best be of service to our communities and the larger Earth community in this time of transition? We can create change one small action at a time, starting with our everyday lives. These actions inspire others, like fractals moving outward and upward.

The single most important action in the heart-centered revolution is to stop pretending that our current sociopolitical systems are in place to serve humanity or the greater Earth community. The most meaningful work in this time is to unplug from the consensus reality and to experience our full range of feelings with the uncertainty, together, and to bring this unapologetic knowing into our collective spaces. Zhiwa Woodbury argues the world is having a #MeToo moment.[1] Time is up on our exploitation of planet Earth. We must act, every day, in each moment, to reflect this reality.

Those of us asleep under the consensus trance require individual and collective rewiring and reprioritization of values, creating the foundations for new worldviews and ways of being. This means that the questions we ask about our beliefs and motivations, the ways we treat others, and the attention we place on our day-to-day actions all matter. These actions create our collective systems, and our systems inform our world.

We must withdraw from our toxic systems, which is often easier said than done. But this is our reminder that participating in BAU enables inaction. We must *slow down* and identify where we can pull out of the dominant paradigm. The systems will cease to exist if enough of us stop believing and participating in them. We can minimize flying and buying material goods that we do not need. We can repair items instead of buying new ones. We can share our resources with one another, grow food for our families and communities, plant native species, and spend our time and energy creating experiences. We can create community response teams so that when an extreme weather event occurs, we have a plan to keep people safe. We can care for community members whose homes have been demolished in a flooding event, a storm, or a wildfire. Through these means, we can free ourselves from the fickle nature of our lawmakers. We can take ownership of our lives and our communities. We can make choices about what creates meaning and joy, and we can start to see change right away.

Will it save the world, right now, today? No. But the actions our planet so desperately needs are not dependent on outcomes. We do the good work because that is what is being asked of us in this critical moment on Earth. We do the work because it aligns with our values and protects what is still here. We also do it because if we do not, we will be perpetuating the same problems and cultural narratives that got us into this predicament in the first place. A focus on meaning, joy, and actions that increase connection, repair, and healing moves us toward wholeness. These actions benefit us, our families, our friends, our community at large and, dare we say, the world. When we take our inner transformation outward, we might see that the world around us changes too. This is *heart-centered activism*, and it is a grassroots movement in the truest sense.

Finding Our Unique Place of Power

LaUra

When I started drafting these steps in my graduate work, it was because I needed a road map to help preserve my sanity as I continued to work to build a better world. My therapists minimized my concerns about the state of the world, many of my friends and family wished I would stop talking about the complex global issues, and I was left with existential despair knowing that so much of what I love and value is being destroyed by the culture of which I am a part. I felt too small to effect any meaningful change. I was a broken twenty-something with years of abuse, neglect, poverty, and loss as my major milestones. How could I muster the courage to keep working for change when power-over systems seemed impenetrable? What could I, as one person, do in the face of so much disruption? How could I minimize suffering and maximize connection?

From the time I started to understand the severity of the predicament, I have been struggling with how to live in a world so misaligned with the fundamental realities of life and life processes. I could not unknow what I know. I had been plucked from the Matrix and felt it was my responsibility, as someone committed to awakening, to create solutions to a solutionless problem. My deep longing told me that if I read more or did enough personal trauma work and organized more groups, then I could help usher in a life-centered existence. But I came to realize that my real power is my ability to sit with the possibility that it might be too late to turn the ship around. To face the possibility head on, with clarity, opens new possibilities for meaning, action, and connection. My power grows because I willingly become acquainted with the profound grief, rage, hopelessness, and despair that cycle within me over and over again. I continue to wake up each day and create community groups to help others process these feelings and explore our next best steps. In an unraveling world, being able to sit with my own feelings and do my own work helps tremendously when I encounter other folks who are also coming to terms with the severity of the crises. Having explored these daunting feelings myself, I'm a better listener. I can connect, relate, and assure the other participants that they are not alone and that this is not a time for solutions but a time for unplugging, getting lost, and embracing the mess. Jung said, "Knowing your own darkness is the best method for dealing with the darkness of other people."[2] I do not fear the darkness in other people because I have become acquainted with my own.

For me, this step is a reminder that I, alone, can't solve the world's problems. I can do my part, play my instrument, and help inspire others to do the work they feel called to do. I remind myself that it is OK for me to take the day off and lie in bed or wade through a river even as the world unravels. I remember how essential it is to unplug from organizing, reading, and thinking critically and to pursue experiences of meaning and joy. This step prompts me time and time again not to take life too seriously, to breathe, and to be with the complexity of growth, decay, and the everyday magic of being alive.

EXPERIENCES, SKILLS, PASSIONS

Heart-centered activism looks different for every person because it is based on creating meaning and joy in service to the larger community and healing. It might be reinvesting energy into conscious parenting so that we are not projecting our fears and insecurities about the state of the world onto our children and into the future. For others, it may look like regenerating landscapes and ecosystems. It could mean planting an organic garden and helping to feed our local community. Or offering classes or childcare. Maybe learning more about Buddhist economics or the transformative justice movement. For some, meaningful action is hosting repair workshops to help mend broken items or sew clothes. Maybe it is meaningful to organize spaces to come together and connect or host weekly dinners or write poetry and sing. Meaningful action could be offering to host rituals and ceremonies for fellow community members. Perhaps meaningful action is engaging with Project Drawdown[3] and implementing real-world climate solutions because every degree of planetary warming matters. Of course, we need holding actions that work to halt the destruction of people, Earth, and life-support systems. We get to choose what types of actions bring us joy, cultivate meaning, and foster reconnection with the world around us.

We have seen members of our support groups go on to do a variety of things: offer support groups of their own, quit their full-time jobs to join direct-action movements, restore the land they're on, change careers, become eco-chaplains, reactivate a law license to offer help at the US-Mexico border, create a documentary on climate grief, and of course become closer with the local community and loved ones. Whatever meaningful action is to each of us, the point of Step 10 is this: when we act from a place of openhearted calm connectedness, we as change makers, healers, and visionaries are most powerful, energized, and resilient.

Some people know right away what types of actions are generative, meaningful, and joy-filled for them. For others, it is not immediately clear where their personal power lies. If finding our personal power is challenging, we can begin by looking at the intersection of our experiences, skills, and passions. All too often we tamp down what is meaningful to us, especially when we have received the message, either explicitly or implicitly, that what is important to us is not valuable to others or is too far on the fringe. But this is the time to do the fringy work that others do not understand. Jim McAuley, a fellow Good Griever, told the group that his heart-centered activism starts with embracing his inner weirdo, which he had previously locked away in his shadow.

OUR INNER WEIRDO IS OUR SUPERPOWER

Aimee

Sometimes, our heart-centered activism doesn't show up in obvious ways. We have to search a little and embrace our inner weirdo. While I've always had a passion for music and making mix CDs (remember those?), it was not until graduate school that I started dabbling in DJing. Between crafting essays, I was looking for a fun, exciting, and safe way to spend time, so I got a small DJ controller. And that was the start of a beautiful relationship full of mixing music and hosting dance parties. Nothing was more satisfying than amping up a room full of people with the perfect playlist.

What does that have to do with climate chaos? On the surface, nothing. But soon after cofounding GGN, I realized that my DJing skills helped prepare me for being a facilitator. DJing is facilitating a party—I stand above the crowd, reading the mood of the room, and adjust my playlist in conversation with the flow of the floor. I start slow and, as the room fills up, raise the energy. If things get a bit too wild, I can dial the tempo back. I have a list of songs in my back pocket for every situation. These lessons translated well to facilitating the 10 Steps. Read the room, stay open to flow, and adjust the energy as necessary.

We are a constant process, changing as we learn, unlearn, and grow. It's important to allow space for our actions to shift as we transform. A variety of action is welcome and needed through the Long Dark, and we never know when or how our unique superpowers will come in handy. And each and every one of us is needed, right now. As we release our inner weirdo from our shadow, we can get to work. No need to wait until we are perfect or are guaranteed an outcome to get started.

EXERCISE: FINDING YOUR SUPERPOWER

Create a list of your experiences, your passions, and your skills. Try to make this list as exhaustive as possible and continue adding to it as time goes on.

Experience: We do not just mean the years of experience one might list on a résumé. What are the lived experiences that have shaped your unique worldview: your geographic location, ethnicity and culture, traumas, peak moments, education (formal or informal), milestones, key conversations, relationships. What are the big and little moments that brought you to this moment in your life? What instances have defined you as a person? Was it a loss? Gaining a degree? Was it a conversation you had?

Skills: Our skill set is made up of the abilities each of us brings to the world. Skill is not necessarily the same as talent. Some skills take a lifetime to develop, but we do not have to be an expert in a skill for it to be meaningful or valuable to the larger world. What are you good at? Were you born with special talents? Have you cultivated them over time?

Passion: What makes you come alive? What do you have a deep yearning to do? What do you wake up in the morning thinking about? What types of actions bring you meaning and joy? What can you get lost in for hours? What do you feel incomplete without? No passion is too frivolous for the heart-centered revolution.

Now review your lists. Do you notice any patterns? Is there overlap between the categories? Which of these actions contribute to healing, connection, and growth? How might your experiences, skills, and passions combine into meaningful and joyful actions in a transitioning world?

Consider how you might bring your experiences, skills, and passions alive in your day-to-day life. In what ways can you share your superpowers with your larger community? This exercise can be ongoing with regular assessment as our inner and outer worlds change.

EXPANDING ACTIVISM

Activism is commonly known as engaging in actions that help bring about social transformation. As previously discussed, we often think of holding actions as conventional activism. However, we cannot only focus on things the dominant culture must stop doing, we need room for what we want to cultivate and grow. Imagining new paradigms and taking steps to build them are also activism. So many types of actions are needed right now, some of which we cannot even imagine yet. This means everyone is invited into activism. Everyone is welcome into the change as we broaden our ideas of conventional activism and engage in actions that are meaningful and generative (see the figure below for some suggestions). As our world reorganizes, we have an opportunity to look to our superpowers and imagine how they can be utilized in service of healing, growth, and connection.

We suggest committing to three meaningful actions for yourself:

1. What is one action you can practice today?
2. What is one action you can practice in the next month?
3. What is one action that will unfold over the next year?

offer childcare help to your community
join or create communal housing
mend & repair items
envision new economic & political systems
deconstruct cultural narratives that perpetuate systems of oppression
embrace your inner weirdo
seek out beauty
take breaks from news & social media
feel your whole range of feelings
spend time with the more-than-human world
cultivate personal boundaries
establish regular grounding practices
personal & collective trauma healing
nourish & move your body
explore & heal your triggers
rewild yourself
play in the dirt
create art & music
reinvest in parenting
take breaks & rest
participate in nonviolent civil disobedience
listen to your intuition
redistribute personal & generational wealth
hold community grief rituals
grow a garden
play
imagine

A few suggested meaningful actions. (Illustration by Sarah Jornsay-Silverberg.)

Community and Emergent Solutions

> Without inner change, there can be no outer change. Without collective change, no change matters.
>
> —Rev. angel Kyodo williams

While the heart-centered revolution begins within, we need systemic change to alter the course of our dying biosphere. We simply cannot create new systems that are life-centered when we are isolated. We need larger systems of individual players interacting with each other to create change. And that means it is on us to bring our distinct power and authentic, vulnerable, messy selves to our communities and share our gifts.

In community, no one individual has to know all the answers or have everything figured out. It is extremely limiting to believe we must solve the world's problems by ourselves. Plus, we may recoil under that type of pressure. We can rely on group consciousness to help illuminate the pathways forward. For example, I am not particularly knowledgeable about economics. While I know we must transition away from capitalism, I am not sure what a new economy would look like—and that is OK. There are others in my community with a passion for economics and degrowth who are imagining new ways of being. I will bring facilitation skills, and someone else will add insight into economics. Yet another person can bring a deep knowledge of council decision-making processes, and someone else will want to discuss a transformative justice approach of accountability. We can remove the pressure to be the individual savior and level up as a collective.

Throughout these steps, we have turned to adrienne maree brown's emergent strategy principles for inspiration. When we follow these principles, they can ripen into *emergent solutions*—which, put simply, are new ways of being that emerge from the powerful synergy of heart-centered activists coming together. As we each bring our passions, skills, and experiences together, we build on one another, spark new ideas and inspiration, and innovate new ways of being that were previously unfathomable to us. Community offers a place for the spiraling up of nuanced thoughts and creative pathways that simply cannot happen when we are alone. Together is the only way forward; we belong to each other.

Emergence is not a destination we reach but a process of cultivating relationships, serious healing, and significant repair and then witnessing what unfolds. We are always creating the conditions for something to arise. Through practicing the steps and connecting in community, we have a say in the types of conditions we are nurturing. We can look to ecosystems for an example of emergence. An ecosystem is made up of a variety of species engaging with one another, as well as with water, energy, food sources, and other elements. An ecosystem never announces, "I have emerged." It is a constant process of growth and contraction, with bursts of vibrancy. We know from ecosystem science that diversity is a strength in any community, and this carries over to our human systems: diversity in experiences, perspectives, ideas, and capabilities builds strength. The more we can engage with each other in brainstorming and heart-centered visioning, the more we can notice collective openings.

JOURNAL EXERCISE: RADICAL IMAGINING

> The inability to imagine a world in which things are different is evidence only of a poor imagination, not of the impossibility of change.
>
> —Rutger Bregman, *Utopia for Realists*

As systems are rearranging and transforming, imagination is an indispensable component in ushering new ways of being into existence. But if we spend too much of our energy thinking of all the things that are not working in our world, we are not creating much time or space for visioning future pathways. What we pay attention to grows. In this exercise, we tap into our imaginations to envision the world we would like to create. The best part is that we do not have to know how to get there.

In your journal, write down answers to the questions below. Do not get bogged down in trying to figure out exactly how to make all these ideas come to life. Do not bat an eye if it seems impractical or too idealistic. If you would like to see that animals receive the same rights as people or free health care for everyone, you do not have to map out how it will be done. This is an exercise in imagining new ways of being, not problem-solving.

This exercise is even more inspiring when we take it outside of our individual journals and share our answers with others, as each person's unique skill set, passions, and experiences have something different to bring to the new paradigms. Build on one another's ideas. You might find some of them are doable today.

- What is invited into the new paradigms? Many of the qualities we want to invite in may exist or have existed in other cultures or in past versions of our own. Let us invite them back in or imagine new ways of being together in these times.

  Example: In the new paradigm, we will operate with council circles for major decisions. We will all learn musical instruments. We will share stories and brew our own beer.
- What is no longer accepted in the new paradigms? What are we leaving behind in the current paradigm? What is not working for us? What is creating walls instead of bridges?

  Example: In the new paradigm, we are leaving behind national boundaries, nuclear weapons, and fossil-fuel energies. We are also leaving behind militarized police.
- What is one character trait within you that you want to see passed on to the new paradigms?

Example: I admire my integrity. In the new paradigm, integrity will be a value that will be encouraged and prioritized.

Parenting through the Predicament

A main focus for GGN over the past few years has been holding affinity groups for parents (or caregivers). We know our young people are suffering. In the years between 2009 and 2018, suicide rates rose 287 percent for individuals under eighteen.[4] They are born into a world full of crises and day after day see the adults around them fail to take radical action to preserve a livable future. Of course the youth are suffering.

In a world that promises more tumult, it can be difficult helping young ones navigate this time. The difficulty increases if we are not caring for our own emotional responses to the predicament. Parents (and adults who have a parent-like roles) play an important part in this time of great transition. We think one of the most important meaningful actions is doing our own work so that we can model healthy relationships with uncertainty, death, and our inner worlds. We can teach younger generations to look for openings and to be absolutely present in the current moment. To do this, we must learn how to feel our feelings and not project our fears and anxieties onto our children. As parents, we can model a life with increased emotional intelligence, a willingness to question our questions, an ability to be with each other without rushing to fix, and a commitment to bringing more meaning, joy, and connection into our lives. We can journey with our children through the Long Dark with resilience, courage, and connection. Good Griever Teddy Kellam said,

> When my children were younger, I wanted their wide-eyed, openhearted selves to build resilience through wonder and feeling the goodness of the world. Nowadays they are teenagers who are regularly exposed to news on the climate crisis, so I cannot honestly say, "Everything will be okay." I remind my teens that while I

don't know how the climate emergency will shake out, there will be so many good things in their lives—falling in love, good music for dancing, foot rubs—all the joyful things we do. When I see that they are struggling, I ask questions and I sit quietly with their pain and anger, rather than rushing to talk them out of it. I don't do it perfectly, and I apologize when I'm not proud of how I've handled things. Over time, as I learn to tolerate the fears that move through me, I feel more powerfully resilient. My children sense the spaciousness inside of me, and they can settle into it, even when the outside world doesn't feel safe.

Park Guthrie, another Good Griever, shared how running through the program helped him practice vital skills that he passes along to his children: "For kids to thrive, they're going to need more than an ability to fight for a livable future. They will need to connect and to create community. To become experts at metabolizing grief while acknowledging and managing loss. To celebrate beauty in the midst of collapse." Responsible adults can help with this, but we must be willing to do our work, too.

The Next Best Step

> What the world needs most is openness: open hearts, open doors, open eyes, open minds, open ears, open souls.
>
> —Robert Muller

We have seen the world change multiple times in the past few years: a pandemic, racial reckonings, an attempted coup on the United States government, the Russian invasion of Ukraine, and the increasing severity of the impacts of climate catastrophe. We do not know what tomorrow will bring, much less the next decade. This is not a reason for inaction. The final emergent strategy principle that adrienne maree brown offers is that "there is always enough time for the right work."[5]

Participating in the heart-centered revolution can feel overwhelming in times of unraveling. The complex predicament is so large, and a single person is limited. The magnitude of loss and uncertainty can feel crushing. Yet, at every twist and turn, we have the opportunity to choose whether we submit to the constriction, disconnection, and fear leading us into resignation and causing more harm, or we act in ways that bring us more alive and connected.

Some people scoff at the steps we have presented here. If we are committed to 1.5 degrees of warming and tipping points have been breached, what does any of this matter? The "we're fucked, let's throw our hands up" attitude exemplifies the isolation and fragmentation of the dominant systems. It is only one worldview's response to the predicament. Believing it is too late and failing to act because of that belief is an expression of powerlessness, hopelessness, helplessness, fear, and unmetabolized grief. It means the dominant culture has done its job by exhausting us and eroding our values and commitments to each other.

We can do better. And we must do better. Discounting this work as "too idealistic" sounds a lot like "it is too hard" or "it is hopeless." When things seem impossible, we make magic together. We set intentions and quiet ourselves, listening to our intuition. We draw down our armor and connect at the heart level. We pay close attention to openings. We remember that life is the greatest mystery, and we have never been in control anyhow. We plant seeds deep within the seed bank, hoping a few will emerge when the conditions are just right. Uncertainty is a pathless opportunity, a powerful disruptor to help us change all aspects of society with a focus on justice, compassion, and relationships. And if it is too late and humans go extinct, at least we have gone out with beauty, awe, and love. The other option? Allowing the current paradigm to continue BAU until everything is obliterated.

There is no "right way" to be in these times. When we weigh all our options, it comes down to the next best step. Which step brings us closer to our personal and collective values? Which direction allows us to create more connection, healing, and growth? What is the next step that allows us to cultivate meaning, seek beauty, and experience joy? What brings us more alive? It becomes easier to refocus our attention from the massive global predicament to our local communities, where we are being asked to exhibit courage and feel around in the dark for the next best step.

There is no single pathway forward. In the song "No Way as the Way," dead prez reminds us that our ways do not have to look the same, and that is OK. (Take a reading pause and go listen to that song.) We are living in a complex time where what we know is continuously thrown into questioning and rearrangement. It is enough in a broken world to forge our path, make mistakes, and be fully alive in ways that cultivate expansion in times of global contraction. We have all the tools we need to transition to a better livelihood for humanity and many of Earth's other inhabitants. The only thing we lack is the commitment and willingness to dream it into reality. Meaning, joy, connection, and courage can serve as our guides as long as there is life.

We must plant seeds, even if we do not get to see them sprout.

Outro and Onward

> When you're in the middle of a big adventure, you don't have time to decide whether you're hopeful or hopeless; all your energy should be right here, in the moment.
>
> —Joanna Macy, *The Wisdom to Survive: Climate Change, Capitalism and Community*

The Great Unraveling has brought us to this moment in human history. As we descend deeper into the Long Dark, a time of awakenings, reckonings, and rearranging, we are forced to change—because change is coming whether we like it or not. We have choices about how we respond to these changes. What will guide us? How will we be with each other? How can we utilize compassion and cultivate connection? Change is the fundamental truth of reality, and much of our current ways of being need to be restructured to live in right relationship with change. As Grace Lee Boggs said, "The time has come for us to reimagine everything."[1] To be alive now is to be in transition, to remove dependence on outcome and instead learn how to dance while falling. This also means that there is infinite room to do meaningful work.

Our journey through each of the 10 Steps is a great place to get started. They are a springboard into the heart-centered revolution that is being birthed. During our support group meetings, we often refer to the steps in the program as a spiral. They are nonlinear, overlapping and intertwining, and, most importantly, we never really "finish" cycling through them. Like a spiral, the steps and what we can learn from them only keep expanding. There are many participants who are GGN regulars, returning to the program again and again to work through the new layers of understanding that unfold each time the steps are revisited. Additionally, showing up to the steps changes with who we practice with. Each member brings unique experiences, passions, and skills, and each and every group is a revolutionary experience where new ideas, insights, and inspiration emerge. Practice groups are containers for unlimited fertility, helping us sow seeds for new ways of being.

So, congratulations, dear one—you are done, but not really. The work will never be complete. We must remain committed to decolonizing our somas, finding openings, and continuously cultivating connections at all levels over fear, constricting, and Othering.

There are no conclusions to the collective crises we face or to the process of grieving them. After we have truly sat with the painful realities of our times—that the climate crisis is now a factor in everything we do, that coloniality runs deep within each of us living in the dominant paradigm, and that we can never truly "save" the world—there comes a deep, tremoring sadness that we must venture through, time and time again. We are being called to practice our highest level of courage—to face the predicament, continuously process our heavy and painful feelings, and find respite in our connections. We do not have to do this work alone. *We cannot do this work alone.* And if we want a chance at creating a just and life-centered future, we must start on the individual level and branch out.

This is not a time to rush to solution making. It is a time to question everything, to help the rearranging process, to endure discomfort and honor mortality, and to radically imagine new ways of being and help bring them into the world. The work is in the journey. It is about the process, not the destination. Becoming, not arriving. The work is to be fully alive in the midst of a crisis but not at the expense of other lives. We live in the process by developing strong communities and following the wisdom provided by the 10 Steps. We explore our pain for the world without shutting down to it. Attempts to avoid the pain deny us opportunities to ripen into mature human beings. The only way out is through.

Eventually, we can arrive at a sense of freedom. We accept that, for better or worse, some systems will inevitably, unpreventably fail. They already are. And they have to, to protect what is left of our biological systems. When we face that reality head on, we can let go of our illusions of control. We focus our attention on the present moment. What is still here? What is worth preserving? How can I be of use in this time and place? What is important right now, and how can I show up fully?

We never completely heal our planetary grief or other heavy or painful feelings. We will continue to be visited by immense pain and suffering because of the state of the world. But we can choose to show up, openheartedly in each moment, and increase connection with others, with the more-than-human world, and with ourselves despite it.

Seeds for the Future

> The one who plants trees, knowing that he or she will never sit in their shade, has at least started to understand the meaning of life.
>
> —Rabindranath Tagore

As an outro to the 10 Steps, we would like to leave you with a few more seeds, invitations we practice embodying each day. May the processes and principles in this book continue to unfold and ripple out in their own ways as they reach into your somas and transform them.

RECONNECTING ON ALL LEVELS

Many of us feel isolated and alone in these times. The unprecedented changes to our biosphere and the cultural transitions that are occurring remind us of the importance of building authentic connections as an antidote to collapsing individually. We heal the radical wound of our perceived separation from the more-than-human world and those we consider Other. We open ourselves to the vulnerability of community. We create space for voices that have been silenced to be heard. And we reconnect with all aspects of ourselves, including our shadow.

May we all feel belonging and connection.

HEALING OUR TRAUMAS AND OPENING TO LOVE

Navigating life can be a traumatic experience. By its very nature, the dominant culture oppresses and exploits. A life can be full of a combination of unmet expectations, abuses, and failures. Being alive while everything we know unravels is the most urgent calling to start healing our individual and collective traumas. The harder we hold on to our wounding, the more easily triggered we will be by suffering and the degradation of life-supporting systems. As the traumas continue, the impact has a compounding effect. A refusal to heal causes more harm as we act out our survival strategies. We courageously attend to the pain of the world without further projecting our own pain. To be present with our transitioning loved one—our sick planet—without our own agenda . . . this is love.

May we each heal those parts of ourselves that carry deep wounding.

FEELING MORE DEEPLY

Most of us do what we can to run from feelings like rage, despair, or grief. But we know that we cannot selectively mute our emotions. If we run from the painful ones, we dampen the pleasurable ones. As we become willing to feel our whole range of emotions, we open up, many of us for the first time, to our full human experience. We are willing to feel both sadness and joy more deeply and, in the process, create stronger, more meaningful, and more loving connections with ourselves and others.

May we exercise patience with ourselves as we continuously process our feelings and embrace the pleasurable ones alongside the painful ones.

LETTING GO OF OUTCOMES

The climate catastrophe is already unfolding in unpredictable ways, and our future is uncertain. We cannot anticipate what the Long Dark will present us with; there is no paved path. In a dynamic world, it serves us to take a note from the Buddha and live in the present moment, detaching from outcomes. We get quiet so that we can listen to our intuition and take the next best step. We learn to cultivate humility and nimbleness. We rest in the magic of the unknown. We cultivate daily practices to help anchor us to the present moment, to each other, and/or to a spiritual practice.

May we all proceed with nimbleness and openness.

DECONSTRUCTING HARMFUL STORIES

The dominant paradigm and its power-over structure are full of stories, norms, and expectations that keep us isolated, disconnected, and perpetuating harm. As we notice how they live in us, we work to deconstruct them. In doing so, we are mindful to minimize harm as we rewrite stories, norms, and expectations that move us toward more connectivity, growth, and healing.

May we all have the courage to assess and deconstruct harmful cultural stories.

EMBRACING COMPLEXITY

We undo our cultural stories that have us categorizing everything we know into binaries and dualisms. We embrace both/and thinking and understand that our perception and biases prevent us from seeing the whole picture. Once we have humbled ourselves, we become more comfortable in the mess, in the discomfort. And we open to the complexity of these times.

May we all embrace complexity in the ways we communicate, think, live, and be with each other.

MENDING AND REPAIR

We mend all we can on an individual and collective level. We focus on repairing relationships and material goods. Being a part of a larger community helps us simplify our lives and lifestyles.

May we learn to mend, repair, and share with community members.

SLOWING DOWN

Our time is precious. We regularly ask ourselves, Are we spending more time working or busying ourselves than we have to? Do we have the ability to minimize work time and create more space for rest, engaging our imagination, relationship building, and doing things that bring us more alive? Can we create more space in our lives to experience meaning and joy? What are the activities that increase our aliveness and move us closer to wholeness?

May we slow down and spend our precious time in ways that are generative, meaningful, and joyful.

QUESTIONING OUR QUESTIONS

We continuously question what we think we know. We have permission to be wrong and we show compassion when others are wrong. As we peel back our conditioning layer by layer, we create space for new (and ancient) ways of knowing to emerge.

May we all cultivate humility and compassion as we embrace new information and ways of knowing.

LOOKING FOR OPENINGS

We create the conditions for a future world that is truly life-centered. Increasing connectivity, prioritizing diversity, and studying the rhythms of life are critical to our next steps. As we take our next steps, we must pause to notice what is bubbling up in the periphery. We must stay with the tension to see what is being birthed in the margins.

May we stick with the tension long enough to notice what's emerging in the periphery.

NURTURING SUPPORTIVE COMMUNITIES

We cannot do this work alone. We connect with others—ideally, locally. We build relationships with our neighbors and rely on one another. We practice boundaries and accountability. We understand that relationships are messy and imperfect but fulfilling and necessary. We create room to be embraced in our messiness and not be canceled. In community, we absolve the need to do everything or know everything. We take breaks and rest. We work together.

May we all build communities of trust, vulnerability, reciprocity, and accountability.

FINDING JOY AMID THE CHAOS

The disruptions through the Long Dark will threaten our sense of groundedness. Yet, there will be moments worth celebrating. We cultivate joy, connection, and meaning amid the losses and disorientation.

May we all learn to dance while falling.

GENERATING MEANINGFUL ACTIONS

A simple formula for making meaning and expressing joy is to locate the intersection where our experiences and skills meet our passions. Taking it a step further, we identify which actions contribute to healing, growth, and connection. Each of us works toward our meaningful actions. We stay open as the world changes around us, acting in service of the larger human and Earth community.

May we all have the courage to pursue generative, meaningful actions.

KEEPING THE VISIONS FLOWING

Imagination, play, and covisioning are key throughout the Long Dark. We, as a species, have never been here before. No maps, no saviors. Each of us requires courage, openness, humility, imagination, and collaboration to guide our pathways forward.

May we create ample space for imagining, visioning, and collaboration.

Onward.

APPENDIX

"BEDROCK" BY KRISTAN KLINGELHOFER

Did you want anything more than a clear blue future for your
children?
Did you eat a steady diet of stories that told you it would all end
okay?
Did you believe in Paris?
Did you get solar panels and backyard chickens and take your kids
camping and
sing before dinner about how you love the earth and will care for
her always?
Did you believe that if you did it all right then it would all be
alright?
I did.

And what happened when your faith began to crack?
What happened when you held a child on your lap and quietly read
the headlines?
The science of methane and bleaching and feedback loops.
Do you remember nursing when you read about tipping points,
afraid your baby might drink your fear?

How do we midwife such hopeful beings into a world that is collapsing
just as they step into it?
Why do we teach them to trust us?

The End of Nature sat on my nightstand but I covered it with a baby blanket.
I thought I should write letters and march and sign petitions but I preferred to
help little hands learn to stack blocks.
My husband raged about it all. He was chipping at my armor and I resented him for it.
I thought it was his fault their childhoods were getting spoiled.

I swung through conversations and through seasons:
hopeful, hopeless, hopeful, hopeless.
Exhausted as a seabird,
trying to land.

My faith was thinning but, still, I held it up like fabric for the children to see.
I wonder now if this was an act of love or of betrayal.

The most brutal moments were the simple questions:
Mommy, why is the sun so red today?
Why do we have to put our favorite things in a bag?
Will the fire burn up our house?
Can I bring all of this?
Why does it hurt to breathe?
Is everything going to be okay, mama?

My voice too high. *Everything is going to be fine, honey!*
Who was I trying to convince, them or myself?
I was desperate to push down the rising panic,
but it always came back at night.
The better my children slept, the worse I did.
They had no idea.

One day, my daughter learned of the mass extinctions.
(*All of them? Forever? Mama . . . really?*)
She still put together an outfit of purple and blue, and
wondered who would sit next to her at lunch.
I saw in her the divide that severed me.
I began to break.
Later, my son read the entire IPCC report in his bedroom. He came
out and said nothing. He didn't even look
at me.
I broke.

I was frozen. I didn't know how to be me anymore.
I didn't know how to parent.
I couldn't pretend, but how could I be honest?

I found Good Grief. The name called to me like a lost relative.
Meeting one, our first moments together, we all shared why we had
come.
We were scared, alone, anxious, depressed, detached, angry, sad.
We each emerged from a cave of isolation. Stepped into the sun of
understanding.
We were crying. We were melting.
Becoming soft again.
I saw the sadness under my husband's rage and I loved him for it.

He felt the fabric of me torn in half and held the pieces in his
quiet hands.

We took off our masks. We all had stories.
The doomers and the activists didn't agree and it didn't matter.
We didn't even talk about the facts. Why would we?
We didn't try to fix each other.
We just listened, hard as we ever had.
Our oldest armor dissolved in the warmth of each other's attention.
Even a joy came, which surprised us. How could this feel good?
Like holding hands in the dark, I suppose.
We learned to need each other the way people used to.

We dug for truth, and here is what we discovered:
under the sedimentary layers of rage, of despair, of a glacial
numbness,
we found a bedrock of grief.
What a strange comfort!
At least grief is sturdy. At least grief is real.
We can stand on this, together.

We kept digging.
Under the grief, we found something deeper:
a hot molten love for this world.
This is what life is made of.
It can burn anything in its path.
We aren't crazy. This love is primordial.
No wonder the losses hurt so much.

And now?

We return to our lives,
secure on that bedrock of grief,
sensing, always, the fluid pulse underneath.
They are the same, really, aren't they?
Hot and cold rock, grief a form of love.

And when the fires come again, as they will,
I will kneel down and lock eyes with any child,
yours or mine.
"I know it is scary, honey. I am scared too.
Here, take my hand."

The impossibilities that used to rip us,
the unanswerable questions of our time,
now we can hold them in our widened arms.
We have grown our capacity.
We have "become immense," as Weller asked.
We contain Whitman's multitudes.
Perhaps we are becoming elders and that is what the world needs now more than anything.

And hope is no longer something we measure
with inadequate tools. It is a way of being.
It looks like this:
We plant fruit trees near the sidewalk, we love teenagers, we sing the old movement songs, we notice the sky, we know our neighbors, we tell the truth, we dance.
We rewild any corner of the world still possible,
hoping to rewild ourselves in the process.

And, of course, we grieve.
You don't leave the bedside of your
mother, your lover, or your child.
The Earth is all of these.
We will be the ones with the courage to cry.
We will not pretend, or turn away.
We will save every scrap of life that we can.
And we will honor each loss in the best way we know: with our
attention to it.
It is sacred work, learning to say goodbye.

Whatever the unknowable future holds,
we now have glimpsed how we will face it.
Connected, reliant,
doing the gritty work of re-imagining.
Sending roots into the rock.

Kristan is a lead GGN facilitator, holding 10-Step circles for parents worldwide, and she helps train facilitators. She is the mother of three children. She believes that our distress stems from our love of the world and that healing begins in community.

GLOSSARY

both/and thinking: Removing binarized thinking and instead embracing more complexity and paradox.

changes in consciousness: One of three types of actions Joanna Macy teaches about that are necessary in these times. These actions promote the deep understanding of interconnectedness.

collective gaslighting: The intentional denial campaign created and paid for by corporations and upheld by politicians to sow doubt about the climate, ecological, and social justice crises.

creating new structures: One of three types of actions Joanna Macy teaches about that are necessary in these times. New structures are essential for creating new life-supporting systems.

death anxiety: The deep fear of our mortality that we suppress because the knowledge is almost too much to bear as a self-reflexive, meaning-making species. Our underlying fear of death shapes many of our unconscious behaviors until we consciously bring to mind our deaths.

decolonizing: Deconstructing Eurocentric worldviews and the impacts of those worldviews.

distress intolerance: The inability to be with our heavy or painful emotions and our desperate urge to avoid or escape them as they arise.

dominant paradigm/culture: A monolithic and homogenized culture characterized by capitalism (more specifically neoliberalism), patriarchy, racism, imperialism, and colonialism. While prevalent in the Western world, this way of being can be found worldwide.

embodied knowledge centers: The head, heart, and gut brains, each creating and sharing its own information and all connected through the vagus nerve.

emergence: A phenomenon within systems theory that observes the ways that parts of a complex system interact to create patterns that would not otherwise exist with the parts alone.

emergent strategy: adrienne maree brown's theory for the ways humans practice complexity and grow the future through relatively simple interactions. It includes methods for cocreating life-centered paradigms.

fast thinking: Daniel Kahneman's term for intuitive thought and automatic mental activities of perception and memory.

fractal: Infinitely complex patterns repeated across different scales yet remain similar.

Great Unraveling: Popularized by Joanna Macy's work, it speaks to the breakdown of our biological and social systems—ecological, cultural, economic, political, and more.

healer: A general term used for any person who helps bring about healing for people or the community. This can be a therapist, spiritual healer, facilitator, body healer, nutritionist, energy healer, and so on.

heart-centered revolution: A revolution that centers feeling, connectedness, and love in times of disconnection and suffering.

holding actions: One of three types of actions Joanna Macy teaches about that are necessary in these times. They are legal actions and social efforts to slow the destruction of people, species, and Earth systems.

human exceptionalism: The cultural understanding that human beings are superior to all other forms of life and the more-than-human world.

impossible choices: The feeling of stuckness as we seek to make decisions that minimize harm, while living in a paradigm based on exploitation, oppression, and constant degradation of our planet. The inability, due to systemic circumstances, to make a choice that does no harm.

Long Dark: A term used by Francis Weller to describe the liminal space between the crumbling ways of being and new (and ancient) ways of being.

new ways of being: A general term building on Thich Nhat Hanh's notion of "interbeing." These ways of being provide an embodied knowing that existence is relational, generative, and diverse.

openings: Opportunities for thought, actions, and new ways of being to emerge that we may have been too distracted or biased to notice before.

positive disintegration: The process of examining and eliminating old beliefs and values so that we can adopt new ones that benefit our personal and collective growth. This theory was developed by the Polish psychologist Kazimierz Dąbrowski.

power-over structure: A system of power where resources, safety, and decision-making belong to an elite few at the expense of most people and Earth.

power-with structure: A system of power where relationships, cooperation, and collaboration are prioritized. Resources, safety, and decision-making are equitably available.

predicament: The interlinked set of problems that have amassed such complexity they can no longer be solved, only faced and lived with.

projection: The displacement of one's feelings, fears, desires, or shadow aspects onto another, often unconsciously.

radical reconnection: An invitation to heal the radical wound of perceived separation by reestablishing, or remembering, our connectedness with ourselves, others, and the more-than-human world.

shadow: A largely hidden part of every human psyche that is composed of the parts of one's personality that are deemed "unacceptable."

slow thinking: Daniel Kahneman's term for the type of thinking that is deliberate and requires effort.

social engagement system: A mode controlled by the vagus nerve that, when we are feeling safe, maintains bodily homeostasis and enhances our ability to connect with others.

soma: Haines's term for the fusion of one's embodied way of thinking, emotions, actions, and worldview.

synchronicity: A meaningful happenstance of two or more events pointing to something beyond the probability of chance.

system: A complex entity made up of smaller individual parts.

systems theory: The study of relationship between the individual parts of complex systems.

terror management theory: A social and evolutionary psychology theory that suggests being reminded of our own mortality causes us to act in self-preserving ways.

vagus nerve: Also called "the wandering nerve," this nerve starts at the brainstem and wanders its way down through much of our torso. It is the key nerve of focus in polyvagal theory and the primary nerve in our autonomic nervous system.

window of tolerance: A state in which our mind is calm, our nervous system is regulated, and we are able to process stimuli and create connections. The window varies from person to person.

RECOMMENDATIONS FOR DEEPER DIVES

We have a living resource list that is regularly updated available at www.goodgriefnetwork.org/livingresourcelist.

Additionally, we have a carefully curated booklist for each of the 10 Steps and overarching themes presented in this book, which can be found at https://bookshop.org/shop/GoodGriefNetwork.

DJ eXis10shAL (a.k.a. Aimee) crafted a Spotify playlist to accompany your reading journey. Find it at www.goodgriefnetwork.org/howtoliveplaylist.

NOTES

PREFACE

1. A head nod to the Dark Mountain Project and the "Uncivilisation" manifesto. Available at https://dark-mountain.net/product/uncivilisation-the-dark-mountain-manifesto/.
2. Columbian Mailman School of Public Health (website), "Depression Is on the Rise in the US, Especially among Young Teens," October 30, 2017, https://www.publichealth.columbia.edu/public-health-now/news/depression-rise-us-especially-among-young-teens; Brandon H. Hidaka, "Depression as a Disease of Modernity: Explanations for Increasing Prevalence," *Journal of Affective Disorders* 140, no. 3 (2012): 205–14, https://doi.org/10.1016/j.jad.2011.12.036.
3. Nabil Ahmed et al., "Inequality Kills: The Unparalleled Action Needed to Combat Unprecedented Inequality in the Wake of COVID-19," Oxfam, January 17, 2022, https://doi.org/10.21201/2022.8465.
4. Amy Goodman and Denis Moynihan, Democracy Now. "Greta Thunberg: Change Is Coming, Whether You Like It or Not." September 19, 2019, https://www.democracynow.org/2019/9/19/greta_thunberg_change_is_coming_whether_.
5. As scholar Farhana Sultana states in her paper, "The Unbearable Heaviness of Climate Coloniality," *Political Geography*, 2022, www.sciencedirect.com/science/article/pii/S096262982200052X, while BIPOC is a contested acronym, it is a commonly used "umbrella signifier of common experiences of colonial and imperial racializations by nonwhite peoples." We, too, will use this acronym, while understanding it has limitations.

6. "Detroit Activist, Philosopher Grace Lee Boggs: 'The Only Way to Survive Is by Taking Care of One Another,'" *Democracy Now!*, April 2, 2010, https://www.democracynow.org/2010/4/2/grace_lee_boggs.

INTRODUCTION

1. We prefer the term *more-than-human world* over the commonly used term the *natural world*. The former reminds us that we are not separate from nature but part of it, just like trees, rocks, water, fungi, and animals. This term was coined by David Abram, and the suggestion to use this term was brought to us by GGN's executive director, Sarah Jornsay-Silverberg.
2. Rob Hopkins, *The Power of Just Doing Stuff: How Local Action Can Change The World* (Cambridge, UK: Green Books, 2013), 45.
3. The author, therapist, and somatic abolitionist Resmaa Menakem uses the term *white-body supremacy*, emphasizing that the white body is the standard against which all other bodies are compared.
4. Francis Weller and Alnoor Ladha, "Deschooling Dialogues: On Initiation, Trauma and Ritual with Francis Weller," *Kosmos*, Winter 2021, https://www.kosmosjournal.org/kj_article/deschooling-dialogues-on-initiation-trauma-and-ritual-with-francis-weller/.
5. Weller and Ladha, "Deschooling Dialogues."
6. Weller and Ladha, "Deschooling Dialogues."
7. We have also seen a backlash against this much-needed awakening as conservative politicians within the United States government push through anticritical race theory laws, preventing public school children from learning about the true history of this country.
8. Sarah Ross, "Resourcing, Pendulation and Titration: Practices for Somatic Experiencing," https://sarahrossphd.com/resourcing-pendulation-titration-practices-somatic-experiencing/.

WE BELONG TO EACH OTHER: THOUGHTS ON COMMUNITY

1. adrienne maree brown, *Emergent Strategy: Shaping Change, Changing Worlds* (Chico, CA: AK Press, 2017), 42.
2. Transition Network website: https://transitionnetwork.org/; Cooperation Jackson website: https://cooperationjackson.org/; Barcelona en Comú website: https://barcelonaencomu.cat/.

3. For more information on the biomimicry movement visit https://biomimicry.org/what-is-biomimicry/. To learn more about permaculture visit the Permaculture Research Institute's website: https://www.permaculturenews.org/what-is-permaculture/
4. Benji Jones, "Indigenous People Are the World's Biggest Conservationists, but They Rarely Get Credit for It," Vox, June 11, 2021, https://www.vox.com/22518592/indigenous-people-conserve-nature-icca.
5. The educators and agents of change Brian Arao and Kristi Clemens first discussed the concept of "brave spaces" in chapter 8 in *The Art of Effective Facilitation: Reflections from Social Justice Educators* (Sterling, VA: Stylus Publishing, 2013).
6. brown, *Emergent Strategy*, 41. She cites this idea as coming from Taj James.
7. Elizabeth Marks et al., "Young People's Voices on Climate Anxiety, Government Betrayal and Moral Injury: A Global Phenomenon," Preprints with *The Lancet*, September 7, 2021, http://dx.doi.org/10.2139/ssrn.3918955.
8. S. Clayton et al., *Mental Health and Our Changing Climate: Impacts, Inequities, Responses* (Washington, DC: American Psychological Association and ecoAmerica, 2021).
9. Andrea Stanley, "The Coming Age of Climate Trauma," *Washington Post Magazine*, October 27, 2021, https://www.washingtonpost.com/magazine/2021/10/27/camp-fire-ptsd/.
10. See efforts made by Climate Psychology Alliance (in the United Kingdom and North America), Climate Psychiatry Alliance, Psychologists for a Safe Climate in Australia, and the Climate Psychology Certificate through the California Institute of Integral Studies.

STEP 1: ACCEPT THE SEVERITY OF THE PREDICAMENT

1. Staci K. Haines, *The Politics of Trauma: Somatics, Healing, and Social Justice* (Berkeley, CA: North Atlantic Books, 2019), 55.
2. Geoffrey Supran and Naomi Oreskes, "Rhetoric and Frame Analysis of ExxonMobil's Climate Change Communications," *One Earth* 4, no. 5 (May 21, 2021): 696–719, https://www.sciencedirect.com/science/article/pii/S2590332221002335; Jessica Corbett, "Secret Footage of ExxonMobil Lobbyists Sparks Calls for Congressional Action," Truthout, July 1, 2021, https://truthout.org/articles/exxonmobil-lobbyists/.

3. Mark Kaufman, "The Carbon Footprint Sham: A 'Successful, Deceptive' PR Campaign," Mashable, accessed June 16, 2022, https://mashable.com/feature/carbon-footprint-pr-campaign-sham/.
4. The idea of consensus trance and consensus reality, along with the direct quote, came to us through Terry Patten's wonderful book *A New Republic of the Heart: An Ethos for Revolutionaries* (Berkeley, CA: North Atlantic Books, 2018), 39.
5. Rachel Nall, "What Are the Long-Term Effects of Gaslighting?," Medical News Today, June 29, 2020, https://www.medicalnewstoday.com/articles/long-term-effects-of-gaslighting#summary29.
6. Panu Pihkala, "Toward a Taxonomy of Climate Emotions" in *Frontiers in Climate*, 3 (January 14, 2022), https://www.doi.org/10.3389/fclim.2021.738154; Glenn A. Albrecht, *Earth Emotions: New Words for a New World* (Ithaca, NY: Cornell University Press, 2019).
7. Vanessa Machado de Oliveira, *Hospicing Modernity: Facing Humanity's Wrongs and the Implications for Social Activism* (Berkeley, CA: North Atlantic Books), 30.
8. Machado de Oliveira, *Hospicing Modernity*, 30.
9. "The Pandemic Is a Portal," YouTube video, streamed live by Haymarket Books on April 23, 2020, 1:35:07, https://www.youtube.com/watch?v=QmQLTnK4QTA.
10. We will not focus on planetary boundaries or climate tipping points in this book, but if you do not yet know about them, viewing the Netflix documentary *Breaking Boundaries: The Science of Our Planet* is a great place to start learning about them. Damian Carrington, environmental editor for the *Guardian*, wrote an informative article on tipping points: "World on Brink of Five 'Disastrous' Climate Tipping Points, Study Finds," published September 8, 2022, and available at https://www.theguardian.com/environment/2022/sep/08/world-on-brink-five-climate-tipping-points-study-finds.
11. IPCC (website), "Summary for Policymakers of IPCC Special Report on Global Warming of 1.5°C Approved by Governments," October 8, 2018, https://www.ipcc.ch/2018/10/08/summary-for-policymakers-of-ipcc-special-report-on-global-warming-of-1-5c-approved-by-governments/.
12. Prentis Hemphill, "Hope, Questioning, and Getting Lost with Bayo Akomolafe," May 3, 2021, in *Finding Our Way*, produced by Eddie Hemphill

and devon de Leña, podcast, season 2, ep. 3, https://www.findingourwaypodcast.com/individual-episodes/s2e3.

13. The roots of the planetary predicament are composed of systemic and compounding issues based in disconnection from the more-than-human world, leading to human exceptionalism and further disconnecting us through to white supremacy. Adding to this, our dependence on logic discounts other ways of knowing and has created a collective naivety and ignorance permitting the submission to the consensus reality without much questioning. To address the climate crisis, we start by addressing the interlinking crises together from their roots. All other measures toward solution-making will fail. We will spend the rest of this book coaxing out these connections.
14. Bayo Akomolafe, "A Slower Urgency: We Will Dance with Mountains," BayoAkomolafe.net, accessed February 2, 2022, https://www.bayoakomolafe.net/post/a-slower-urgency.
15. Resmaa Menakem, *My Grandmother's Hands: Racialized Trauma and the Pathway to Mending Our Hearts and Bodies* (Las Vegas: Central Recovery Press, 2017), 20.
16. Haines, *Politics of Trauma*, 36.
17. Emily Nagoski, *Come as You Are: The Surprising New Science That Will Transform Your Sex Life* (New York: Simon & Schuster, 2015), 115.
18. Nagoski, *Come as You Are*, 114.
19. Daniel J. Siegel, *Pocket Guide to Interpersonal Neurobiology* (New York: W. W. Norton & Company Ltd., 2012).
20. Emily Nagoski, "Complete the Stress Cycle," *Experience Life*, September 6, 2021, https://experiencelife.lifetime.life/article/complete-the-stress-cycle/.
21. Brené Brown has a large body of work, and we recommend digging into it. One great entry point into her feelings work is her book *The Gifts of Imperfection: Let Go of Who You Think You Are Supposed to Be and Embrace Who You Are* (Center City, MN : Hazelden, 2010). Another is her TED Talk, "The Power of Vulnerability," available on YouTube: https://www.youtube.com/watch?v=iCvmsMzlF7o&ab_channel=TED.
22. Karin Nila et al., "Mindfulness-Based Stress Reduction (MBSR) Enhances Distress Tolerance and Resilience through Changes in Mindfulness," *Mental Health and Prevention* 4, no. 1 (2016): 36–41; Shannon

Sauer-Zavala et al., "The Role of Negative Affectivity and Negative Reactivity to Emotions in Predicting Outcomes in the Unified Protocol for the Transdiagnostic Treatment of Emotional Disorders," *Behaviour Research and Therapy* 50, no. 9 (2012): 551–57.

23. From Dr. Albert Wong's website: "This is an adaptation of Mike Bostock's adaptation of Geoffrey Roberts's Emotion Wheel using D3's partition layout. Robert's 2015 work appears to be based on a vocabulary wheel by Kaitlin Robbs from 2014, which in turn appears to be based on The Feeling Wheel published by Gloria Willcox in 1982." To view the actual feelings wheel on the website and for more information go to www.dralbertwong.com/feelings-wheel/.
24. In the talk at the University of Utah in the fall of 2013, Moser drew from a chapter she authored: Susanne C. Moser, "Getting Real about It: Meeting the Psychological and Social Demands of a World in Distress," SusanneMoser.com, accessed February 2, 2022, http://www.susannemoser.com/documents/Moser_Getting_Real_About_It-preprint.pdf.

STEP 2: BE WITH UNCERTAINTY

1. United Nations (website), "Famine Knocking at the Door of 41 Million Worldwide, WFP Warns," UN News, June 22, 2021, https://news.un.org/en/story/2021/06/1094472.
2. Tetsuji Ida, "Climate Refugees—the World's Forgotten Victims," World Economic Forum, June 18, 2021, https://www.weforum.org/agenda/2021/06/climate-refugees-the-world-s-forgotten-victims.
3. Farhana Sultana, "The Unbearable Heaviness of Climate Coloniality," *Political Geography*, 2022, https://www.sciencedirect.com/science/article/pii/S096262982200052X.
4. University of British Columbia, "Changes in Earth's Crust Caused Oxygen to Fill the Atmosphere," Phys.org, September 18, 2017, https://phys.org/news/2017-09-earth-crust-oxygen-atmosphere.html.
5. Field Museum, "Mass Extinction 250 Million Years Ago Sparked Dramatic Shift to Complex Marine Ecosystems," ScienceDaily, November 28, 2006, https://www.sciencedaily.com/releases/2006/11/061126121112.htm.
6. Octavia Butler, *Parable of the Sower* (New York: Grand Central Publishing, 1993), 3.

7. Joanna Macy, *World as Lover, World as Self: Courage for Global Justice and Ecological Renewal* (Berkeley, CA: Parallax Press, 2007), 95.
8. Prentis Hemphill, "Hope, Questioning, and Getting Lost with Bayo Akomolafe," May 3, 2021, in *Finding Our Way*, produced by Eddie Hemphill and devon de Leña, podcast, season 2, ep. 3, https://www.findingourwaypodcast.com/individual-episodes/s2e3.
9. Jerry was a teacher in the "Warrior for the Human Spirit" training with Margaret Wheatley in October 2019.
10. Hans-Rudolf Berthoud and Winfried L Neuhuber, "Functional and Chemical Anatomy of the Afferent Vagal System," *Autonomic Neuroscience* 85, no. 1–3 (2000): 1–17, https://doi.org/10.1016/S1566-0702(00)00215-0.
11. Two suggested resources are Stephen Porges, *The Pocket Guide to the Polyvagal Theory: The Transformative Power of Feeling Safe* (New York: W. W. Norton & Company, 2017), and Deb Dana, *Polyvagal Exercises for Safety and Connection: 50 Client-Centered Practices*, Norton Series on Interpersonal Neurobiology (New York: W. W. Norton & Company, 2020).
12. The trauma expert Peter Levine taught the concept of "relative safety" during a digital talk on April 2, 2022, called "Threat of War: Somatic Rebalancing," https://fb.watch/eyjakHVCYm/.
13. Staci K. Haines explains that in generative somatics, an organization that supports social and climate justice movements in radically transforming society, trauma is understood to come from fractures in any or all of these essential qualities of safety, belonging, and dignity.
14. Francis Weller, "Rough Initiations: In the Absence of the Ordinary," *Kosmos*, Winter 2021, https://www.kosmosjournal.org/kj_article/rough-initiations/.
15. For more information on climate trauma, see Zhiwa Woodbury, "Climate Trauma: Toward a New Taxonomy of Trauma," *Ecopsychology* (March 2019): 1–8, https://doi.org/10.1089/eco.2018.0021; Michael Richardson, "Climate Trauma, or the Affects of the Catastrophe to Come," *Environmental Humanities* 10, no. 1 (May 2, 2018): 1–19, https://doi.org/10.1215/22011919-4385444; and Charlie Hertzog Young, "Diagnosing Climate Trauma," *Ecologist* (November 4, 2021), https://theecologist.org/2021/nov/04/diagnosing-climate-trauma.

16. Box adapted from the Trauma Practice (website), "Types of Trauma," accessed February 2, 2022, https://traumapractice.co.uk/types-of-trauma, and Zhiwa Woodbury, "Climate Trauma: Towards a New Taxonomy of Traumatology," *Ecopsychology* 11, no. 1 (March 2019): 1–8, https://doi.org/10.1089/eco.2018.0021.
17. Staci K. Haines, *The Politics of Trauma: Somatics, Healing, and Social Justice* (Berkeley, CA: North Atlantic Books, 2019), 39.
18. Marilyn Morgan, "Neuroscience and Psychotherapy," *Hakomi Forum*, no. 16–17 (Summer 2006): 9–22, https://hakomiinstitute.com/Forum/Issue16-17/2_NeurosciTherap-Format.pdf.
19. Cheryl Richardson, *Take Time for Your Life: A 7-Step Program for Creating the Life You Want* (New York: Harmony, 1999). The quote is also available on the author's website: https://cherylrichardson.com/newsletters/week-41-letter-wish-send-every-checked-healthcare-provider/.
20. Pema Chödrön, *When Things Fall Apart: Heart Advice for Difficult Times* (Boston: Shambhala Publications, 1997), 71–72.
21. Terry Patten, *A New Republic of the Heart: An Ethos for Revolutionaries* (Berkeley, CA: North Atlantic Books, 2018), 139.
22. Patten, *A New Republic*, 157–59.
23. George F. Young et al., "Starling Flock Networks Manage Uncertainty in Consensus at Low Cost," *PLOS*, January 31, 2013, https://doi.org/10.1371/journal.pcbi.1002894.
24. adrienne maree brown, *Emergent Strategy: Shaping Change, Changing Worlds* (Chico, CA: AK Press, 2017), 41.
25. Fractal Foundation, "What Are Fractals?," accessed June 17, 2022, https://fractalfoundation.org/resources/what-are-fractals/.
26. brown, *Emergent Strategy*, 41.
27. James Baldwin, *The Devil Finds Work: Essays* (New York: Vintage International, 1976), 32.

STEP 3: HONOR MY MORTALITY AND THE MORTALITY OF ALL

1. Learn more at the website for Earth Overshoot Day, https://www.overshootday.org/.
2. Many countries in the Global North exploit countries from the Global South, extracting and shipping their local resources to faraway lands. Most often, the profits do not go back to the Global South, and these

communities are also left with the environmental toxins produced from extraction.

3. "Country Overshoot Days," Earth Overshoot Day, accessed June 17, 2022, https://www.overshootday.org/newsroom/country-overshoot-days/.
4. Catie Gould, "How I Learned to Cope with Climate Grief," Bike Portland, March 5, 2020, https://bikeportland.org/2020/03/05/how-i-learned-to-cope-with-climate-grief-311902.
5. Roy Scranton, *Learning How to Die in the Anthropocene: Reflections on the End of a Civilization* (San Francisco: City Lights Books, 2015).
6. Roy Scranton, "Learning How to Die in the Anthropocene," *Opinionator* (blog), *New York Times*, November 10, 2013, https://opinionator.blogs.nytimes.com/2013/11/10/learning-how-to-die-in-the-anthropocene/.
7. Ernest Becker, *Denial of Death* (New York: Free Press, 1973).
8. Psychology Today Staff, "Terror Management Theory," *Psychology Today*, accessed June 17, 2022, https://www.psychologytoday.com/intl/basics/terror-management-theory.
9. Cathy Cox and Jamie Arndt, "The Theory," Terror Management Theory, last updated February 1, 2008, https://tmt.missouri.edu/index.html.
10. A large collection of terror management theory studies can be found at https://tmt.missouri.edu/publications.html.
11. You should read it. It is eerily accurate concerning what we have lived through the past few years. J. L. Dickinson, "The People Paradox: Self-Esteem Striving, Immortality Ideologies, and Human Response to Climate Change," *Ecology and Society* 14, 1 (2009): art. 34, http://www.ecologyandsociety.org/vol14/iss1/art34/.
12. Janis L. Dickinson, "Why Climate Change Threatens Our Inner Life and Survival," *Cornell Chronicle*, January 21, 2010, https://news.cornell.edu/stories/2010/01/climate-change-threatens-our-inner-and-outer-lives.
13. Katia Patin, "The Rise of Eco-Fascism," Coda Story, January 19, 2021, https://www.codastory.com/waronscience/the-rise-of-eco-fascism/.
14. Leigh Phillips, "Brown Shirts, Green Dreams," *Noēma*, June 23, 2022, https://www.noemamag.com/brown-shirts-green-dreams. For further information on ecofascism and white supremacy in the ecological movement, see the work of Sarah Jaquette Ray at https://sarahjaquetteray.com.
15. Oeschger Centre for Climate Change Research, "The Climate Is Warming Faster Than It Has in the Last 2,000 Years," University of Bern

(website), July 24, 2019, https://www.oeschger.unibe.ch/about_us/news/warm_period/index_eng.html.

16. John P. Rafferty, "Biodiversity Loss," *Encyclopedia Britannica*, June 14, 2019, https://www.britannica.com/science/biodiversity-loss#ref1266690.
17. adrienne maree brown, *Emergent Strategy: Shaping Change, Changing Worlds* (Chico, CA: AK Press, 2017), 42.
18. This story was taught to us by our friend and colleague Bonita Ford in her book *Embers of Hope*. Check it out. To read a translated version of the story from the *Sallatha Sutta* visit https://www.accesstoinsight.org/tipitaka/sn/sn36/sn36.006.nypo.html.
19. This was inspired by an exercise taught to us by Andrea Bernstein in the Munay-Ki tradition.

STEP 4: DO INNER WORK

1. Susan Moser, personal communication, March 14, 2014.
2. Thomas Moore, *Care of the Soul: A Guide for Cultivating Depth and Sacredness in Everyday Life* (New York: HarperPerennial, 1992).
3. Vanessa Machado de Oliveira, *Hospicing Modernity: Facing Humanity's Wrongs and the Implications for Social Activism* (Berkeley, CA: North Atlantic Books, 2021), 20.
4. Center for Sustainable Systems, "US Cities Factsheet." University of Michigan, 2020, pub. no. CSS09-06, http://css.umich.edu/factsheets/us-cities-factsheet.
5. Instagram post by Aubrey Marcus, May 16, 2021, https://www.instagram.com/p/CO8SoN5nHOq/.
6. Emma Betuel, "Scientists Discover a Major Lasting Benefit of Growing Up outside the City," Pocket, accessed June 18, 2022, https://getpocket.com/explore/item/scientists-discover-a-major-lasting-benefit-of-growing-up-outside-the-city?utm_source=pocket-newtab.
7. Joanne B. Newbury et al., "Association of Air Pollution Exposure with Psychotic Experiences during Adolescence," *JAMA Psychiatry* 76, no. 6 (2019): 614–23, https://doi.org/10.1001/jamapsychiatry.2019.0056.
8. John Marzluff et al., eds., *An Introduction to Urban Ecology as an Interaction between Humans and Nature* (New York: Springer, 2008), https://doi.org/10.1007/978-0-387-73412-5.

9. Harvard Health Publishing, "A Prescription for Better Health: Go Alfresco," Harvard University, October 12, 2010, https://www.health.harvard.edu/newsletter_article/a-prescription-for-better-health-go-alfresco.
10. Cornell University, "Spending Time in Nature Reduces Stress," ScienceDaily, February 25, 2020, https://www.sciencedaily.com/releases/2020/02/200225164210.htm.
11. Gregory N. Bratman et al., "Nature Experience Reduces Rumination and Subgenual Prefrontal Cortex Activation," *Proceedings of the National Academy of Sciences* 112, no. 28 (June 2015), https://doi.org/10.1073/pnas.1510459112.
12. Beth Shaw, "When Trauma Gets Stuck in the Body," Psychology Today, October 23, 2019, https://www.psychologytoday.com/us/blog/in-the-body/201910/when-trauma-gets-stuck-in-the-body.

STEP 5: DEVELOP AWARENESS OF BIASES AND PERCEPTION

1. adrienne maree brown, *Emergent Strategy: Shaping Change, Changing Worlds* (Chico, CA: AK Press, 2017), 41.
2. *Encyclopedia Britannica Online*, s.v. "physiology," accessed June 20, 2022, https://www.britannica.com/science/information-theory/Physiology.
3. Colin G. Scanes, "Animal Perception Including Differences with Humans," chapter 1 in *Animals and Human Society*, ed. Colin G. Scanes and Samia R. Toukhsati (London: Academic, 2018), 1–11, https://doi.org/10.1016/B978-0-12-805247-1.00001-0.
4. Scanes, chapter 1.
5. Daniel Kahneman, *Thinking, Fast and Slow* (New York: Farrar, Straus and Giroux, 2011).
6. Tara Kadioglu, "Why Slow Thinking Wins," *Boston Globe*, July 26, 2015, https://www.bostonglobe.com/ideas/2015/07/25/the-power-slow-thinking/ToZbzYl7rGoyVMCtsZ7WnJ/story.html.
7. Christopher Hitchens, *Letters to a Young Contrarian* (New York: Basic Books, 2001), 102.
8. Rohini Radhakrishnan, "How Many Organs Are There in the Body?," MedicineNet, March 16, 2021, https://www.medicinenet.com/how_many_organs_are_there_in_the_body/article.htm.
9. Hans-Rudolf Berthoud and Winfried L Neuhuber, "Functional and Chem-

ical Anatomy of the Afferent Vagal System," *Autonomic Neuroscience* 85, no. 1–3 (2000): 1–17, https://doi.org/10.1016/S1566-0702(00)00215-0.

10. Michael Mosley, "The Second Brain in Our Stomachs," BBC, July 11, 2012, https://www.bbc.com/news/health-18779997.
11. Heather Gerrie, "Our Second Brain: More Than a Gut Feeling," University of British Columbia, Graduate Program in Neuroscience, accessed June 20, 2022, https://neuroscience.centreforbrainhealth.ca/our-second-brain-more-gut-feeling.
12. Ying Li and Chung Owyang, "Musings on the Wanderer: What's New in Our Understanding of Vago-Vagal Reflexes? V. Remodeling of Vagus and Enteric Neural Circuitry after Vagal Injury," *American Journal of Physiology—Gastrointestinal and Liver Physiology* 285, no. 3 (September 2003): G461–469, https://doi.org/10.1152/ajpgi.00119.2003.
13. Rebecca Seal, "Unlocking the 'Gut Microbiome'—and Its Massive Significance to Our Health," *Guardian*, July 11, 2021, https://www.theguardian.com/society/2021/jul/11/unlocking-the-gut-microbiome-and-its-massive-significance-to-our-health.
14. Gerrie, "Our Second Brain."
15. Rollin McCraty, *The Science of the Heart: Exploring the Role of the Heart in Human Performance*, vol. 2 (Boulder Creek, CA: HeartMath Institute, 2015).
16. Ruth Feldman, et al., "Mother and Infant Coordinate Heart Rhythms through Episodes of Interaction Synchrony," *Infant Behavior and Development* 34, no. 4 (December 2011): 569–77, https://doi.org/10.1016/j.infbeh.2011.06.008.
17. J. A. Mikels et al., "Should I Go with My Gut? Investigating the Benefits of Emotion-Focused Decision Making," *Emotion* 11, no. 4 (2011): 743–53, https://doi.org/10.1037/a0023986.
18. Tina Fossella, "Human Nature, Buddha Nature: An Interview with John Welwood," *Tricycle*, Spring 2011, https://www.tricycle.org/magazine/human-nature-buddha-nature/.
19. White-bodied people have been appropriating the traditions of Earth's Indigenous wisdom keepers for a long time. This is a reminder that if you are white-bodied, it is your job to listen, learn, and amplify the voices of Indigenous wisdom keepers and their traditions with permission.

20. PenelopePenguin, "Doomscrolling," Urban Dictionary, March 24, 2020, https://www.urbandictionary.com/define.php?term=doomscrolling.
21. *The Social Dilemma*, directed by Jeff Orlowski-Yang, Netflix 2020.

STEP 6: PRACTICE GRATITUDE, SEEK BEAUTY, AND CREATE CONNECTIONS

1. Prathik Kini et al., "The Effects of Gratitude Expression on Neural Activity," *NeuroImage* 128 (March 2016): 1–10, https://www.sciencedirect.com/science/article/abs/pii/S1053811915011532.
2. Glenn Fox, "What Science Reveals about Gratitude's Impact on the Brain," Mindful.org, June 10, 2019, https://www.mindful.org/what-the-brain-reveals-about-gratitude/.
3. Chris Jordan, "Can Beauty Save the World?," TEDxSeattle, accessed June 21, 2022, https://www.ted.com/talks/chris_jordan_can_beauty_save_our_planet.
4. Anjuli Sherin, *Joyous Resilience: A Path to Individual Healing and Collective Thriving in an Inequitable World* (Berkeley, CA: North Atlantic Books, 2021), 9.
5. Morton Schatzman, "Obituary: Viktor Frankl," *Independent*, September 5, 1997, https://www.independent.co.uk/news/people/obituary-viktor-frankl-1237506.html.
6. Viktor E. Frankl, *Man's Search for Meaning* (Boston: Beacon Press, 2006), 106.
7. *Cambridge Dictionary*, s.v. "magic," accessed August 11, 2021, https://dictionary.cambridge.org/dictionary/english/magic.
8. Carl G. Jung, "Synchronicity," in *Jung on Synchronicity and the Paranormal*, ed. R. Main (London: Taylor & Francis, 2005), 91–98.

STEP 7: TAKE BREAKS AND REST

1. Lenora E. Houseworth, "The Radical History of Self-Care," *Teen Vogue*, January 14, 2021, https://www.teenvogue.com/story/the-radical-history-of-self-care.
2. Aimaloghi Eromosele, "There Is No Self-Care without Community Care," Urge, November 10, 2020, https://urge.org/there-is-no-self-care-without-community-care/.

3. Nick Chiles, "8 Black Panther Party Programs That Were More Empowering Than Federal Government Programs," *Atlanta Black Star*, March 26, 2015, https://atlantablackstar.com/2015/03/26/8-black-panther-party-programs-that-were-more-empowering-than-federal-government-programs/4/.
4. "Radical Self Care: Angela Davis," YouTube video posted by Afropunk, December 17, 2018, 4:27, https://www.youtube.com/watch?v=Q1cHoL4vaBs.
5. "Black Panther Greatest Threat to US Security," *Desert Sun* 42, no. 296 (July 16, 1969), https://cdnc.ucr.edu/?a=d&d=DS19690716.2.89&e=-------en--20--1--txt-txIN--------1.
6. A wonderful introduction to Audre Lorde's work is Audre Lorde, *The Selected Works of Audre Lorde* (New York: W. W. Norton & Company, 2020).
7. The Nap Ministry (@thenapministry), Instagram post, October 12, 2020, https://www.instagram.com/p/CGQSrBVlOb4/.
8. Saundra Dalton-Smith, *Sacred Rest: Recover Your Life, Renew Your Energy, Restore Your Sanity* (New York: Faithwords, 2017).
9. Goran Medic, Micheline Wille, and Michiel E. H. Hemels, "Short- and Long-Term Health Consequences of Sleep Disruption," *Nature and Science of Sleep* 9 (2017): 151–61, https://doi.org/10.2147/NSS.S134864.
10. Dalton-Smith, *Sacred Rest*, 36.
11. Molly Shea-Shine, "Still Stressed after a Long Break? You Might Not Be Taking the Right Type of Rest," *Fast Company*, September 4, 2019, https://www.fastcompany.com/90398474/how-to-take-breaks-that-are-right-for-you.
12. Terry Ward, "7 Strategies for Truly Restorative Rest," CNN, updated January 24, 2021, https://edition.cnn.com/2021/01/24/health/restorative-sleep-strategies-wellness/index.html.
13. Dalton-Smith, *Sacred Rest*, 97.
14. This table was adapted from Dr. Saundra Dalton-Smith's interview with CNN. See note 20 in this chapter.
15. Sara W. Lazar et al., "Meditation Experience Is Associated with Increased Cortical Thickness," *Neuroreport* 16, no. 17 (2005): 1893–97, https://doi.org/10.1097/01.wnr.0000186598.66243.19.
16. Roberto Grujičić, "Prefrontal Cortex," Ken Hub, last reviewed March 20, 2022, www.kenhub.com/en/library/anatomy/prefrontal-cortex.

17. Britta K. Hölzel et al., "Mindfulness Practice Leads to Increases in Regional Brain Gray Matter Density," *Psychiatry Research* 191, no. 1 (2011): 36–43, https://doi.org/10.1016/j.pscychresns.2010.08.006.
18. Tammi R. A. Kral et al., "Impact of Short- and Long-Term Mindfulness Meditation Training on Amygdala Reactivity to Emotional Stimuli," *NeuroImage* 181 (2018): 301–13, https://doi.org/10.1016/j.neuroimage.2018.07.013.
19. Judson A. Brewer et al., "Meditation Experience Is Associated with Differences in Default Mode Network Activity and Connectivity," *Proceedings of the National Academy of Sciences* 108, no. 50 (December 2011): 20254–59, https://doi.org/10.1073/pnas.1112029108.
20. Zameena Mejia, "10 Science-Backed Benefits of Meditation" *Forbes*, February 23, 2022, https://www.forbes.com/health/mind/benefits-of-meditation/
21. David Treleaven, "Trauma-Sensitive Mindfulness," accessed August 2, 2021, https://www.davidtreleaven.com/.
22. For our science-minded readers, we know the "funnel" in titration is a burette, and the "valve" is a stopcock. We have simplified for ease of the practice.

STEP 8: GRIEVE THE HARM I HAVE CAUSED

1. For years, Nestlé Waters North America has stolen water from Michigan and sold it back to us at an astronomical profit at the expense of our groundwater. They operate the brands Ice Mountain, Poland Spring, Deer Park, Ozarka, Zephyrhills, Arrowhead, Pure Life, and Splash. The company has taken "more than 1 million gallons per day at its Ice Mountain bottled water operation in Mecosta and Osceola counties—for nothing more than a $200-per-year state permit." Keith Matheny, "Nestlé Bottled Water Operations in Michigan Sold as Part of $4.3B Deal to NY Private Equity Firm," *Detroit Free Press*, February 2, 2018, https://www.freep.com/story/news/local/michigan/2021/02/18/nestle-bottled-water-sold-one-rock-metropoulos-equity-ice-mountain/4497448001/.
2. Do not even get us started: Laura Sullivan, "How Big Oil Misled the Public into Believing Plastic Would Be Recycled," NPR, September 11, 2020, https://www.npr.org/2020/09/11/897692090/how-big-oil-misled-the-public-into-believing-plastic-would-be-recycled.

3. An Apprenticeship with Sorrow course with Francis Weller, online, Summer 2020.
4. Joan Sutherland, "Here at the End of the World," *Lion's Roar*, April 7, 2020, https://www.lionsroar.com/here-at-the-end-of-the-world/.
5. Sutherland, "Here at the End."
6. As of July 19, 2022, Bezos's net worth was estimated at $31,227,309,235.43. "Jeff Bezos' Net Worth Today Is . . ." Available at https://playback.fm/jeff-bezos-net-worth, accessed July 20, 2022.
7. Vanessa Machado de Oliveira, *Hospicing Modernity: Facing Humanity's Wrongs and the Implications for Social Activism* (Berkeley, CA: North Atlantic Books, 2021), 20.
8. Staci K. Haines, *The Politics of Trauma: Somatics, Healing, and Social Justice* (Berkeley, CA: North Atlantic Books, 2019), 55.
9. We strongly recommend Resmaa Menakem's book, *My Grandmother's Hands: Racialized Trauma and the Pathway to Mending Our Hearts and Bodies* (Las Vegas: Central Recovery Press, 2017).
10. Menakem uses "bodies of culture," where other theorists or teachers use "bodies of color."
11. "Resmaa Menakem on Why Healing Racism Begins with the Body," University of Arizona, Center for Compassion Studies, accessed June 21, 2022, https://compassioncenter.arizona.edu/podcast/resmaa-menakem.
12. Resmaa Menakem has developed a program called "somatic abolitionism," where he offers different practices for individuals and collectives depending on whether you exist in a body of culture or you are white-bodied (www.resmaa.com/movement).
13. This practice was adapted from Mindful.org. You can find their version of the practice at www.mindful.org/just-like-me-compassion-practice/.
14. Haines, *Politics of Trauma*, 23.
15. Resmaa Menakem, "Unlocking the Genius of Your Body," *Psychology Today*, December 18, 2020, https://www.psychologytoday.com/us/blog/somatic-abolitionism/202012/unlocking-the-genius-your-body.
16. Joanna Macy, *A Wild Love for the World: And the Work of Our Time* (Boulder: Shambhala, 2020), 298. We are guessing that if you are reading this book right now that you are a Shambhala warrior. To watch Joanna Macy explain the full prophecy, head here: https://vimeo.com/191169785.
17. adrienne maree brown, *Emergent Strategy: Shaping Change, Changing*

Worlds (Chico CA: AK Press, 2017), 42. This idea is Mervyn Marcano's riff off Stephen Covey's work, as cited in her book.

18. adrienne maree brown, *We Will Not Cancel Us: And Other Dreams of Transformative Justice*, Chico CA: AK Press, 2020), 11.
19. Mia Mingus, "Transformative Justice: A Brief Description," https://transformharm.org/transformative-justice-a-brief-description/.
20. Mingus, "Transformative Justice."

STEP 9: SHOW UP

1. Brené Brown, *The Gifts of Imperfection: Let Go of Who You Think You Are Supposed to Be and Embrace Who You Are* (Center City, MN: Hazelden, 2010),13.
2. Francis Weller and Alnoor Ladha, "Deschooling Dialogues: On Initiation, Trauma and Ritual with Francis Weller," *Kosmos*, Winter 2021, https://www.kosmosjournal.org/kj_article/deschooling-dialogues-on-initiation-trauma-and-ritual-with-francis-weller/.
3. Brené Brown, "The Power of Vulnerability," TED Talk, https://www.youtube.com/watch?v=iCvmsMzlF7o&ab_channel=TED.
4. Brené Brown defines wholeheartedness in *Gifts of Imperfection: Let Go of Who You Think You're Supposed to Be and Embrace Who You Are*: "Wholehearted living is about engaging in our lives from a place of worthiness. It means cultivating the courage, compassion, and connection to wake up in the morning and think, *No matter what gets done and how much is left undone, I am enough.* It's going to bed at night thinking, *Yes, I am imperfect and vulnerable and sometimes afraid but that doesn't change the truth that I am worthy of love and belonging.*" Brown has a wholehearted inventory that you can take to assess your strengths and opportunities for growth: https://brenebrown.com/wholeheartedinventory/.
5. A head nod to the brilliant poet Mary Oliver.
6. Staci K. Haines, *The Politics of Trauma: Somatics, Healing, and Social Justice* (Berkeley, CA: North Atlantic Books, 2019), 164.
7. Haines, *Politics of Trauma*, 164.
8. adrienne maree brown, *Pleasure Activism: The Politics of Feeling Good* (Chico, CA: AK Press, 2019), 15.
9. The three rules of dysfunctional families, from Dr. Claudia Black in *It Could Never Happen to Me: Growing Up with Addiction as Youngsters,*

Adolescents, and Adults (Center City, MN: Hazelden, 2002), are widely used in Adult Children of Alcoholics spaces.

10. adrienne maree brown, *Emergent Strategy: Shaping Change, Changing Worlds* (Chico CA: AK Press, 2017), 42.

STEP 10: REINVEST IN MEANINGFUL EFFORTS

1. Zhiwa Woodbury, "Mother Earth's #MeToo! Moment," *Panpsychology NOW!*, August 6, 2019, https://ecopsychologynow.blog/2019/08/06/mother-earths-metoo-moment/.
2. Carl G. Jung, *C. G. Jung Letters, Volume 1* (Princeton, NJ: Princeton University Press, 1992), 237.
3. See https://www.drawdown.org/.
4. E. Brignone et al., "Trends in the Diagnosis of Diseases of Despair in the United States, 2009–2018: A Retrospective Cohort Study," *BMJ Open* 10, no. 10 (2020), https://doi.org/10.1136/bmjopen-2020-037679.
5. adrienne maree brown, *Emergent Strategy: Shaping Change, Changing Worlds* (Chico CA: AK Press, 2017), 41.

OUTRO AND ONWARD

1. Grace Lee Boggs, "Reimagine Everything," *Race, Poverty & the Environment* 19, no. 2 (2012): 44–45, http://www.jstor.org/stable/41806667.

ACKNOWLEDGMENTS

It has been a very long and excruciating journey to birth this book. It took everything we had. Honestly, it took more than we had. And nevertheless, here it is—our words are in the world only because so many believed in us and the work we put our blood, sweat, and tears into. We would need many more pages to personally acknowledge everyone who helped us get here. GGN exists because of the encouragement and support of so many. If you are not personally mentioned, please know we hold great gratitude for you and tried to keep this section as concise as possible. Some people listed below belong in many categories, but may only be mentioned once for brevity.

First, we would like to thank Joanna Macy for paving a heart-centered path for so many of us who were adrift.

We would also like to thank those whose bodies of work provided the spark that grounds much of our work: Bayo Akomolafe, Carolyn Baker, adrienne maree brown, Brené Brown, Staci K. Haines, Bill McKibben, Susi Moser, Krista Tippet, Francis Weller, Rev. angel Kyodo williams, and Terry Tempest Williams.

We honor our ancestors whose guidance, determination, and perseverance help us meet the moment.

Thank you to our team who helped this book come to life: our agent, Rebecca Gradinger, and our editors: Sara Bercholz, Emily Coughlin, and Peter Schumacher. We extend our gratitude to the Shambhala design and production team for making the book beautiful, and to our marketing and publicity team for helping us share these words with the world. An extra special thank you to Jill Rogers for your attentiveness and close read.

Chelsie Rivera, thank you for your hours spent listening to us ramble on and on and for giving the book its backbone, structure, and direction. This book wouldn't exist without you.

Thank you to our earliest participants, patrons, friends, colleagues, and longtime supporters (in no particular order): Leah Hogsten, Vaughn Lovejoy, Victoria Emerson, Amy O'Conner, Ray Wheeler, Sophie Hayes, Sarah Patrick, Kiera Bitter, Phillip Jakobsberg, Dana Snell, Cecile Chaix, Sophie Howarth, Vanessa Barrington, Erin Geesaman Rabke, Roger Ross, Ashley Soltysiak, William Rivers, Barbara Gilbert, Jessie Baines, Bill Grabill, Eric Garza, Tiffany Rolston, Auberon Wolf, Rachel Malena-Chan, Josephine Tucker, Yasmin Ellis, Josh Swenson, Nicholas D'Orazio, Stefanie Valovic, Jen Seelig, Brooke Willis, Sarina Leete, Lisa Hodges, Richard Pauli, Bobbie Mooney, Ken Winter, Kavita Pillay, Tera Atwood, Nathan Braendle, Emilie Jordao, Amanda Lloyd, Kacy Suther, Leon Silverberg, Andy Miller, Barbara Baron, Christopher Matthias, Terry Dance-Bennink, Britt Wray, Trebbe Johnson, Brooke McNamara, Michael Dowd, Ayrel Clark-Proffitt, Roscoe Emrys Grey, Katie Flint, Dianna VanderDoes, Laura Johnson, Monica Prince, Erin Olschewski, Blair Nelsen, Abbey Koshak, Katie Goldman, and Stinne Storm.

We extend gratitude to each person we had the honor of sitting with in a 10-Step circle. Thank you for showing up.

Sarah Jornsay-Silverberg, thank you for your heart and your dedication to this work and for taking turns driving the boat. Dean LaCoe, thank you for your early involvement in the Network and for providing lifeboats along the way. We extend gratitude to Adair Kovac and Steve St. Peter for facilitating the program in its early stages with all the bumps and bruises.

Thank you, Christopher Hormel, for your generosity and belief in our work and words. And, to Mary Migliorelli, your thoughtful gifts and reminders to sustain ourselves have been lifelines. Patrick Gagnon, please thank Zelda for all of the things.

We hold infinite gratitude for Alli Harbertson, Bridget Stuchly, and Brian Emerson for hosting our pilot Good Grief sessions.

For your hearts and your willingness to hold others on this path, we thank the first cohort of FLOW facilitators: Farah Ali, Sarah Birch, LuAnn Collins, Tiffany Felch, Karen Hansen, Andrew Harrell, Subbalakshmi Iyer, Teddy Kellam, Kristan Klingelhofer, Jacqueline Landreth, Lou Leet, Marta Neto, Bradley Pitts, Kasia Stepien, Sarah Stoeckl, Liz Wade, and Rosie Walford.

For GGN's snazzy new logo and colors, we thank Victor Rivera. And for the years of tech help, we are indebted to Michael Rezl.

Finally, we extend our gratitude and love to Cathie and Brian Lewis; Ashlee and Jude Lewis; Jacci, Justin, Mackenzie, Caleb, and Emma Storey; Tammy Henry; and Miles, Romeo, and GusGus for always having our backs.

LaUra

For helping me find the inspiration and ideas to write, I honor and thank the Atlantic and Pacific Oceans, the ancient redwoods in northern California, and the rivers, lakes, and cottonwoods of Scottsbluff, Nebraska.

For my teachers who believe in my voice, even when I do not: Delores Brock, Danielle Marsh Cloutier, Merrie Hammel, Alice Johnson, Gail Turnwald, Bruce Vigneault, Heather Wolf, Anna Monfils, Tracy Galarowicz, Brett Clark, Teresa Cohn, Brooke Williams, and TTW. Thank you to Charles Novitski for taking my spark and fanning it into a flame by teaching the course "Critical Issues Facing Humanity." This book is a continuation of those learnings.

I hold so much love and admiration for the late Jane Fenton, who reminds me still to believe in my magic.

I owe a debt of gratitude to all the families who took me in along my journey, including the Spindlers, Guthries, Peterses, Henrys, Lewises, and Goldmans.

Thank you to the Adult Children of Alcoholics program that taught me how to live and helped inform GGN's 10-Step program. To the variety of healers who have helped light my path, thank you: Kinde Nebeker, Will Gale, Andrea Bernstein, Maggie Hippman, and Lindsey Frischer.

For providing a serene writing space replete with two beautiful pups, a flourishing garden, and nearby access to the ocean and redwoods, I extend gratitude to Sarah Jaquette Ray and her family.

Thank you, Lynne Schmidt, for checking in during tough times.

Thank you also to Dick Meyer, Sarah Bartlett, and Maria Hernandez for the encouragement and support.

For their love and lessons, I honor Brittain, Ollie, Kara, Kellie, Sarah, Lyn, and Kay.

Aimee

I am grateful for my grandparents: Shirley, Steve, Mildred, Doyle, Elaine, and Bob.

I owe much to my teachers: lisa eddy, Dr. Rochelle Harris, Jeffery Bean, Robert Fanning, Matt Roberson, Dr. Steve Bailey, Marya Hornbacher, Sister Pat Schnapp, Dr. Miah Arnold, Laura Newbern, Terry Tempest Williams, Mirabai Starr, Anne Lamott, Rev. angel Kyodo williams, Merlyn Mowrey, and Kinde Nebeker.

Thank you to my healers: Lisa Livingston, Maria Romo, Janine Mayo, Dr. Cathy Rojas, Dr. Mary Webster, Dr. Carolyn Van Cleave, Dr. Sunil Rangwani, Dr. Shanti Mohling, Alethea Jones, Kathleen Rude, Dr. Meghan Roper, and the dedicated team at Polaris Project.

A special note of recognition and gratitude to the Sisters of Mercy at St. Bernardine Home in Ohio, to their loving community of friends, Josie Setzler, and the communicative foxes and trees on the land where they reside.

Last but not least, I bow to the hummingbirds who keep me company, the many rivers who hold my emotions when no one else can, and the maple trees of my homelands whose lifelong friendship continues to be a lifeline.

INDEX

ABOUT THE AUTHORS

LaUra Schmidt

LaUra is the founder of the Good Grief Network and the brain behind the 10 Steps to Resilience & Empowerment in a Chaotic Climate program and the FLOW Facilitation Training modality. She is a truth seeker, community builder, cultural critic, trainer, and facilitator. As the granddaughter and grandniece of holocaust survivors, LaUra has long been captivated by the human condition. She is a lifelong student, curator, and practitioner of personal and collective resilience strategies. Inspiration finds her in natural landscapes and honest, openhearted dialogue.

LaUra grew up in rural Michigan and graduated with a BS in environmental studies, biology, and religious studies from Central Michigan University. Her MS is in environmental humanities from the University of Utah. She holds certificates in Integrative Somatic Trauma Therapy and Climate Psychology. When LaUra isn't watching nature documentaries or eating burritos, she is adventuring with her two naughty rescue pups, Gus and Romeo.

Aimee Lewis Reau

Aimee is the cofounder of the Good Grief Network and the heart behind the 10 Steps to Resilience & Empowerment in a Chaotic Climate program and the FLOW Facilitation Training modality.

She was born and raised in Adrian, Michigan, and is the proud auntie of four amazing niece-phews. Aimee is an edgy and reverent contemplative, healer, and yoga/intuitive movement instructor. She maintains a regular dance practice because she learned from Alice Walker, "Hard times require furious dancing." In her free time, she DJs under the name eXis10shAL. Aimee received her bachelor's degree in English, poetry, and religion from Central Michigan University before obtaining her MFA in creative nonfiction from Georgia College and State University.

Chelsie Rivera

Chelsie is a California-based author with roots in small-town Kentucky. She received her MFA in fiction writing from Georgia College and State University.